Climate Crisis

Editor: Danielle Lobban

Volume 426

independence
educational publishers

First published by Independence Educational Publishers

The Studio, High Green

Great Shelford

Cambridge CB22 5EG

England

© Independence 2023

ISBN-13: 978 1 86168 886 6

Printed in Great Britain

Zenith Print Group

Acknowledgements

The publisher is grateful for permission to reproduce the material in this book. While every care has been taken to trace and acknowledge copyright, the publisher tenders its apology for any accidental infringement or where copyright has proved untraceable. The publisher would be pleased to come to a suitable arrangement in any such case with the rightful owner.

The material reproduced in **issues** books is provided as an educational resource only. The views, opinions and information contained within reprinted material in **issues** books do not necessarily represent those of Independence Educational Publishers and its employees.

Images

Cover image courtesy of iStock. All other images courtesy of Freepik, Pixabay and Unsplash.

Additional acknowledgements

With thanks to the Independence team: Shelley Baldry, Tracy Biram, Klaudia Sommer and Jackie Staines.

Danielle Lobban

Cambridge, May 2023

Contents

Introduction

Climate Crisis is Volume 427 in the **issues** series. The aim of the series is to offer current, diverse information about important issues in our world, from a UK perspective.

About Climate Crisis

Climate change and global warming have led to a climate crisis. This book explores the causes, the impact on us and how we can work together to help take care of our planet.

Our sources

Titles in the **issues** series are designed to function as educational resource books, providing a balanced overview of a specific subject.

The information in our books is comprised of facts, articles and opinions from many different sources, including:

- Newspaper reports and opinion pieces
- Website factsheets
- Magazine and journal articles
- Statistics and surveys
- Government reports
- Literature from special interest groups.

A note on critical evaluation

Because the information reprinted here is from a number of different sources, readers should bear in mind the origin of the text and whether the source is likely to have a particular bias when presenting information (or when conducting their research). It is hoped that, as you read about the many aspects of the issues explored in this book, you will critically evaluate the information presented.

It is important that you decide whether you are being presented with facts or opinions. Does the writer give a biased or unbiased report? If an opinion is being expressed, do you agree with the writer? Is there potential bias to the 'facts' or statistics behind an article?

Activities

Throughout this book, you will find a selection of assignments and activities designed to help you engage with the articles you have been reading and to explore your own opinions. Some tasks will take longer than others and there is a mixture of design, writing and research-based activities that you can complete alone or in a group.

Further research

At the end of each article we have listed its source and a website that you can visit if you would like to conduct your own research. Please remember to critically evaluate any sources that you consult and consider whether the information you are viewing is accurate and unbiased.

Issues Online

The **issues** series of books is complemented by our online resource, issuesonline.co.uk

On the Issues Online website you will find a wealth of information, covering over 70 topics, to support the PSHE and RSE curriculum.

Why Issues Online?

Researching a topic? Issues Online is the best place to start for...

Librarians

Issues Online is an essential tool for librarians: feel confident you are signposting safe, reliable, user-friendly online resources to students and teaching staff alike. We provide multi-user concurrent access, so no waiting around for another student to finish with a resource. Issues Online also provides FREE downloadable posters for your shelf/wall/table displays.

Teachers

Issues Online is an ideal resource for lesson planning, inspiring lively debate in class and setting lessons and homework tasks.

Our accessible, engaging content helps deepen students' knowledge, promotes critical thinking and develops independent learning skills.

Issues Online saves precious preparation time. We wade through the wealth of material on the internet to filter the best quality, most relevant and up-to-date information you need to start exploring a topic.

Our carefully selected, balanced content presents an overview and insight into each topic from a variety of sources and viewpoints.

Students

Issues Online is designed to support your studies in a broad range of topics, particularly social issues relevant to young people today.

Thousands of articles, statistics and infographs instantly available to help you with research and assignments.

With 24/7 access using the powerful Algolia search system, you can find relevant information quickly, easily and safely anytime from your laptop, tablet or smartphone, in class or at home.

Visit issuesonline.co.uk to find out more!

Chapter 4

Climate Change & Global Warming

What is climate change?

Climate change refers to long-term shifts in temperatures and weather patterns. These shifts may be natural, such as through variations in the solar cycle. But since the 1800s, human activities have been the main driver of climate change, primarily due to burning fossil fuels like coal, oil and gas.

Burning fossil fuels generates greenhouse gas emissions that act like a blanket wrapped around the Earth, trapping the sun's heat and raising temperatures.

Examples of greenhouse gas emissions that are causing climate change include carbon dioxide and methane. These come from using gasoline for driving a car or coal for heating a building, for example. Clearing land and forests can also release carbon dioxide. Landfills for garbage are a major source of methane emissions. Energy, industry, transport, buildings, agriculture and land use are among the main emitters.

Greenhouse gas concentrations are at their highest levels in 2 million years

And emissions continue to rise. As a result, the Earth is now about 1.1°C warmer than it was in the late 1800s. The last decade (2011-2020) was the warmest on record.

Many people think climate change mainly means warmer temperatures. But temperature rise is only the beginning of the story. Because the Earth is a system, where everything is connected, changes in one area can influence changes in all others.

The consequences of climate change now include, among others, intense droughts, water scarcity, severe fires, rising sea levels, flooding, melting polar ice, catastrophic storms and declining biodiversity.

People are experiencing climate change in diverse ways

Climate change can affect our health, ability to grow food, housing, safety and work. Some of us are already more vulnerable to climate impacts, such as people living in small island nations and other developing countries. Conditions like sea-level rise and saltwater intrusion have advanced to the point where whole communities have had to relocate, and protracted droughts are putting people at risk of famine. In the future, the number of 'climate refugees' is expected to rise.

Every increase in global warming matters

In a series of UN reports, thousands of scientists and government reviewers agreed that limiting global temperature rise to no more than 1.5°C would help us avoid the worst climate impacts and maintain a livable climate. Yet policies currently in place point to a 2.8°C temperature rise by the end of the century.

The emissions that cause climate change come from every part of the world and affect everyone, but some countries produce much more than others. The 100 least-emitting countries generate 3 per cent of total emissions. The 10 countries with the largest emissions contribute 68 per cent. Everyone must take climate action, but people and countries creating more of the problem have a greater responsibility to act first.

We face a huge challenge but already know many solutions

Many climate change solutions can deliver economic benefits while improving our lives and protecting the environment. We also have global frameworks and agreements to guide progress, such as the Sustainable Development Goals, the UN Framework Convention on Climate Change and the Paris Agreement. Three broad categories of action are: cutting emissions, adapting to climate impacts and financing required adjustments.

Switching energy systems from fossil fuels to renewables like solar or wind will reduce the emissions driving climate change. But we have to start right now. While a growing coalition of countries is committing to net zero emissions by 2050, about half of emissions cuts must be in place by 2030 to keep warming below 1.5°C. Fossil fuel production must decline by roughly 6 per cent per year between 2020 and 2030.

Adapting to climate consequences protects people, homes, businesses, livelihoods, infrastructure and natural ecosystems. It covers current impacts and those likely in the future. Adaptation will be required everywhere, but must be prioritised now for the most vulnerable people with the fewest resources to cope with climate hazards. The rate of return can be high. Early warning systems for disasters, for instance, save lives and property, and can deliver benefits up to 10 times the initial cost.

We can pay the bill now, or pay dearly in the future

Climate action requires significant financial investments by governments and businesses. But climate inaction is vastly more expensive. One critical step is for industrialized countries to fulfil their commitment to provide $100 billion a year to developing countries so they can adapt and move towards greener economies.

www.un.org

What is the climate crisis, why is it happening and what are the impacts?

The average global temperature has risen by more than 1C in the last 140 years – what's going on and does it matter?

By Harry Cockburn

Global warming, climate change, climate crisis, climate breakdown, and climate emergency are all terms which describe a major threat to life as we know it on our planet, but what are people talking about when they use these terms?

Climate and weather

The first important aspect to understanding the world's climate is distinguishing 'climate' from 'weather'.

Weather refers only to the changeable short-term atmospheric conditions in a place or area, while climate refers to a much longer pattern of weather behaviour, and can change from season to season.

But there is also the overall climate of the Earth, which is understood through combining the long-term climate records together to build a picture of the trends affecting our planet.

While weather patterns can change over just a few hours, climates usually take hundreds, thousands or even millions of years to change.

However, since the middle of the 20th century, scientists around the world have been warning that climates all around the world, which had been stable for thousands of years, are beginning to change fast and that it is human activity which is largely causing the current changes.

What is happening?

Overall, the world is getting warmer – this is why the process is often known as 'global warming'.

The average global temperature on Earth has increased by more than 1C (2F) since 1880. Two-thirds of that warming has occurred since just 1975, at a rate of roughly 0.15-0.20C per decade.

However the world is not uniformly getting hotter. Some areas, such as the Arctic and Antarctic are warming faster than others, and rising average temperatures can have numerous other consequences, which is why the process is better described as 'climate change' or 'climate breakdown'.

From a human perspective the huge array of negative impacts on us and other species mean it can be thought of as a 'climate crisis' or 'climate emergency'.

How is the climate changing?

The Earth's climate is always changing, and before we examine why scientists believe humans are having such a large impact on the climate, it is important to recognise that various natural factors cause major changes to the climate.

These include:

- The Earth's distance to the sun. This increases and decreases over a 100,000-year cycle, during which the Earth's orbit becomes more elliptical, and then more circular again. Currently the Earth's orbit is at its most circular, and over the coming millenia, each year during the northern hemisphere's summer months, the Earth will reach the furthest point from the sun as our planet conducts its 12-month-long orbits, at which point, less solar radiation reaches Earth. It is believed this cycle – known as the Milankovitch Cycle – is a significant factor in causing ice ages.

- The Sun can also produce different levels of solar radiation. It does this over an 11-year time span. In the past, average temperatures have loosely tracked solar activity (within a small range), but since the 1950s, there has been no net increase in solar radiation, indeed, in recent decades, the Sun's activity has even slightly dropped compared to decades at the end of the 20th century, while average global temperatures have risen markedly.

- The oceans covering 70 per cent of the world's surface also have a significant impact on climate. This big blue wet thing is continuously exchanging heat, carbon and moisture with the atmosphere, while powerful currents move masses of water around the world, which impact local climates, and clouds form over oceans which are highly reflective and can reflect the sun's energy back into space. Changes to the heat and carbon storage capacity of the world's oceans – due to factors including ice cover, sea levels, and carbon dioxide levels – can impact many other climate-affecting processes, such as the oceans' currents, salinity, temperature and the winds which form over the surface.

- Volcanic eruptions are also natural processes which can have major impacts on the climate. When a powerful volcano erupts, huge amounts of sulphur dioxide, dust and ash are ejected high into the atmosphere, which can reflect sunlight and have a temporary cooling effect which can last some years, though it is not permanent.

Human-driven climate change

Records show that over the last 6,000 years the Earth was on a long slow cooling trajectory, which then suddenly ended during the late Victorian period, around 150 years ago. At this time the human world had already embarked upon a global campaign of industrialisation – a process still underway.

The dawn of industrialisation ignited a planet-wide scramble to excavate exceptionally combustible, energy-rich fuels such as coal, oil and gas, to provide the power required for an endless array of processes including manufacturing, transport, heating and cooling buildings, and domestic use.

This fossil fuel extraction boom is still underway. Though many countries are increasingly switching to renewable energy sources, in 2020, fossil fuels still accounted for 84 per cent of global energy generation, and around the world companies are still planning to keep extracting them.

When fossil fuels are burnt, they emit gases which act to trap the Sun's heat within the atmosphere, thereby heating up the planet like a greenhouse. This is why they are known as greenhouse gases.

As the process of industrialisation altered societies and produced breakthroughs, particularly in medicine and agriculture, populations of humans have risen sharply around the world.

As well as growing populations demanding greater energy resources – particularly in the richest societies – more people also require more food. As a result, agriculture has had an increasing effect on land use and greenhouse gas emissions – particularly for meat production.

The key greenhouse gases are:

- Carbon dioxide

- Methane

- Ozone

- Nitrous oxide

- Chlorofluorocarbons

They all come from different processes, have different concentrations in the atmosphere, and have different impacts over different timescales.

Carbon dioxide accounts for about 80 per cent of all greenhouse gases in the atmosphere, but methane, which makes up about 10 per cent, is much more powerful, though over a shorter amount of time.

Nitrous oxide accounts for about 7 per cent, and the fluorinated gases, which are synthetic and very powerful greenhouse gases emitted from a variety of industrial processes, make up the remaining 3 per cent.

Impacts of the climate crisis

A warming world has terrifying consequences. We are already witnessing many of the impacts of climate change, in the form of extreme weather events and an increasing rate of sea level rise, but if the current rate of warming continues, then the worst is still yet to come.

Scientists have warned we will see more frequent and more intense periods of drought; devastating heatwaves which could increase desertification and make areas of the world uninhabitable; more powerful storms; flooding; warming oceans; melting glaciers; retreating sea ice; lack of snow cover; rising sea levels and the destruction of coastal cities where millions live, and biodiversity collapse, risking famines and global food shortages, and therefore increasing risks of conflict and raising the likelihood of mass migrations.

Response

Responding to the multifarious threats posed by the climate crisis is set to be one of our species' biggest challenges to date.

This is because for many societies, a successful response will involve major recalibrations to people's fundamental values and expectations, as well as deep changes to the systems and institutions our societies have built themselves on over the last century or more.

But if we are unequal to this recalibration the terrible repercussions will intensify and loom ahead of us like an increasingly sinister spectre.

Solutions

The one key thing humans must do to slow down our journey into fiery doom is stop producing emissions of greenhouse gases.

The other key thing we must do is repair the damage we have done to the natural world – this will draw the pollutants out of the atmosphere and help slow down the heating process.

Governments around the world are currently only taking small steps to reduce certain emissions and protect and restore a few select ecosystems. These are glimmers of recognition of the problem, but much more needs to be done.

26 July 2022

The scientific basics of climate change

By Jack Miller

Greenhouse gases (GHGs) emitted by human activity (also referred to as 'anthropogenic emissions') have increased substantially since the industrial revolution, particularly in recent decades.

Although the Earth's climate has been evolving for millions of years, research shows it's extremely likely that these emissions have led to global warming in a short space of time. This in turn is causing climate change.

This Insight gives an overview of the basic concepts underlying GHG emissions, global warming and climate change, with references to further information and underlying research. The Insight UK and global emissions and temperature trends provides information on the data discussed here.

What is a greenhouse gas?

The Earth's atmosphere is made up mainly of nitrogen, oxygen and argon (78.1%, 20.9% and 0.934% in dry air, respectively), as well as a mixture of other gases at much lower concentration.

When energy from sunlight reaches the Earth, much of it is absorbed by land or the oceans, heating the surface. Over time, the heated surface releases this energy. Some of it is absorbed by greenhouse gases, preventing it from leaving the atmosphere.

GHGs are released during many day-to-day activities, such as driving petrol cars or heating homes, as well as industrial production. GHGs trap heat in the atmosphere. Carbon dioxide (CO_2) is the most dominant GHG as it is emitted in greatest quantity by human activity – primarily through fossil fuel burning. CO_2 tends to remain in the atmosphere for hundreds of years.

Other GHGs emitted by human activity have a stronger greenhouse effect than CO_2. However, these either naturally break down in the atmosphere more quickly (such as methane, which takes a few decades) or are emitted in small quantities (such as nitrous oxide, and some of the man-made chemicals known as F-gases). While water vapour is the most abundant GHG, it only lasts for a few days before returning to the surface as precipitation.

Human emissions tip the natural balance

Greenhouse gases have always been an integral part of the atmosphere. Although GHGs currently only make up a small fraction (less than 1%, excluding water vapour) they play an important role in retaining heat from the sun and ensuring we have a liveable planet.

Many natural processes absorb or emit CO_2. For example, photosynthesis absorbs CO_2 and decomposing organic matter releases it. In total, roughly the same amount of CO_2 is emitted and removed by natural processes. This 'carbon cycle', in which emissions are balanced by removals, has kept the atmospheric concentration of CO_2 stable for many thousands of years.

However, emissions of CO_2 resulting from human activity (and other GHGs, particularly methane), have risen sharply since the industrial revolution. This tips the carbon cycle out of balance and increases the atmospheric concentration. This is despite the fact that human emissions of CO_2 are approximately 20 times smaller than natural emissions.

What is global warming?

The atmospheric concentration of CO_2 is now at its highest in several million years, and research shows that the heat trapped by this and other GHGs is increasing the average global temperature. This increasing temperature trend is known as global warming.

Other factors such as changes in the Earth's orbit and volcanic eruptions also affect the global climate, but research has shown it is extremely likely these factors alone cannot account for the warming that has taken place, without also considering human effects.

There has been 1.0°C of global warming since pre-industrial times. This is a clear warming trend shown in measurements averaged over decades and geographic areas. Since records began in 1884, all ten of the UK's ten warmest years have occurred since 2002. Globally, the past five years have been the warmest of the last 140 years.

What is climate change?

Climate change refers to the changes in global weather patterns driven by global warming.

Increasing global temperature has widespread effects on natural systems over land and oceans. While weather varies locally from day to day, when looking at long-term trends over large geographical areas, there are statistically noticeable shifts in weather patterns.

Climate change has been observed in patterns of temperature, humidity and rainfall, as well as in the frequency or intensity of extreme weather events. These are complex effects that are influenced by many natural and human factors, making it difficult to precisely predict how they will develop in future.

Warmer air holds more moisture, so rainfall is increasing on average across the planet, but there is a lot of regional variation. Generally, over the course of the 21st century wet areas are projected to become wetter, and dry areas to become drier. This also means that the intensity of droughts and heavy rainfall is likely to increase, along with the impacts on food security and other human systems that this entails.

Temperature limits and global emissions

Under the Paris Agreement, nearly all governments worldwide agreed to limit global warming to between 1.5 and 2.0°C by the end of the century. The Intergovernmental Panel on Climate Change estimates that to limit global warming to 1.5°C, global emissions need to be roughly halved by around 2030 (compared to 2018) and reach net zero around 2050.

Net zero refers to a situation in which emissions reduce to almost zero, and any remaining emissions are removed from the atmosphere. The UK Government has a goal of net zero GHG emissions by 2050.

24 June 2020

Design

Design a poster to explain the difference between climate change and global warming.

What causes climate change?

Burning fossil fuels for energy, transport and industry releases greenhouse gases, which cause global warming. Things like farming, cutting down forests and overfishing are making it worse. There is no doubt that human activities are causing climate change, which means we are also able to stop it.

The main cause of climate change is burning fossil fuels – such as coal, oil and gas – to produce energy and power transport.

Along with other human activities, like cutting down forests and farming, this releases heat-trapping pollution called greenhouse gases into the atmosphere, warming the planet and destabilising the climate.

The world is now warming faster than at any point in human history. The effects are extreme weather like heavy rain or droughts, long-term shifts in weather patterns, melting ice and rising sea levels. Climate change is already having huge impacts on people and the environment worldwide.

Human activities that release greenhouse gases need to be urgently curbed as much as possible to ensure a stable climate and safe world for everyone.

Where do greenhouse gases come from?

There are a number of human activities that we know to be damaging to the climate. These mostly involve generating energy, because huge amounts of energy are needed to keep our modern world running.

- Generating energy – a lot of power generation for electricity and the vast majority of home heating are still done by burning fossil fuels, such as gas. In the UK, emissions from electricity have gone down rapidly in recent years, thanks to our reductions in burning coal for energy and dramatic increases in renewable energy generation.

- Transport – cars, buses, trains, trucks, ships and planes, (unless electric and charged with renewable energy), all produce emissions by burning fossil fuels. In the UK,

transport is the biggest contributor to climate change, responsible for 27% of emissions in 2019, mostly from cars. International aviation and shipping will continue to be a significant contributor to climate change until demand reduces or alternatives to fossil fuels become available.

- Food production – livestock reared for meat and dairy products emit methane, and agricultural soils emit gases like nitrous oxide, which is made from nitrogen in the soil through the use of fertiliser. As food production increases (with more fertilisers, more livestock, and the need for more crops to feed livestock), emissions will also increase.

- Deforestation – because trees store carbon as they grow, cutting or burning down trees releases that carbon into the atmosphere. Farmers may cut down trees or clear land using fire to produce soya for animal feed, such as in the Amazon. In other parts of the world, natural forests are cleared for timber, mining or palm oil.

- Powering industry – since the Industrial Revolution began in the 18th century in the UK, humans have burned fuel such as coal, oil and gas in order to drive large-scale industries. Industrial emissions come from producing things like cement, iron, steel, electronics, plastics and clothing. All countries are now largely dependent on fossil fuels to build and sustain their economies.

- Plastics and waste – plastics are made from fossil fuels, releasing emissions through their production. Globally, about 40% of plastics are used as packaging. Because so little is recycled (and it would be hard to recycle that much plastic anyway), dealing with waste releases emissions when incinerated (burned) or put into landfill – making it a bigger climate problem than it initially seems.

What are greenhouse gases and how do they cause climate change?

Greenhouse gases include carbon dioxide (CO2), methane (CH4) and nitrous oxide (N2O), which trap heat in the Earth's atmosphere, increasing the average temperature worldwide.

These gases are naturally present in the atmosphere but human activities have massively increased them, trapping heat that then causes climate change.

Carbon dioxide
From burning fossil fuels, deforestation

Methane
From natural gas, permafrost melt, flooding

Nitrous oxide
From fertiliser used in farming

How does Greenpeace campaign against the causes of climate change?

Greenpeace campaigns to stop climate change by persuading governments and companies to change their practices. Governments need to make the right policies – including regulation of company practices – that can help the climate, environment and communities recover.

While Greenpeace and its supporters have had many victories over the years, there is still a lot to do to stop climate change. This is especially urgent because of tipping points that may make it difficult or impossible to recover.

The tipping points that could make climate change irreversible

The Intergovernmental Panel on Climate Change has warned that there could be catastrophic consequences if humanity allows global temperatures to warm by over 2°C (the ultimate limit set by the Paris Agreement).

While scientists now agree that atmospheric warming can be stopped once emissions are brought down or stopped, tipping points complicate things.

A climate tipping point is when small changes combine to become significant enough to cause larger, more critical changes to our climate and our planet, which are likely to be irreversible.

Here are the major tipping points scientists are warning governments about:

- Polar ice sheets collapsing in Greenland and Antarctica – while it is melting slowly, the eventual collapse of the Greenland ice sheet would be irreversible, and sea levels around the world would rise by up to seven metres, leaving cities like Miami and Mumbai underwater. Scientists are also now concerned about the potential collapse of the West Antarctic ice sheet, which would also have extreme effects on the coastlines around the world.

- Arctic permafrost melt – as the atmosphere heats up, the Arctic permafrost is melting, releasing greenhouse gases stored underneath it, such as methane.

- Changing oceans are shifting weather patterns – fresh water from the Greenland ice sheet is melting into the Atlantic ocean, causing the Gulf Stream to slow, leading to extreme cold snaps and colder winters in the US and Europe. The oceans are also absorbing heat generated by greenhouse gas emissions, affecting wildlife and livelihoods around the world, including strengthening El Niño and La Niña weather patterns around the Pacific Ocean.

- Amazon rainforest collapse – the Amazon rainforest is being destroyed and burned to make way for farming, and now produces more than a billion tonnes of carbon dioxide a year, which is now more than it absorbs. If it dries up, billions more tonnes of carbon dioxide would be emitted into the atmosphere, disrupting rainfall across South America and altering climate patterns in other parts of the world.

www.greenpeace.org.uk

Climate questions answered

What are the effects of climate change?

The effects of climate change are with us right now. Millions are suffering already. And younger generations are being robbed of their future on a healthy, liveable planet.

What are the solutions to climate change?

Climate change is already an urgent threat to millions of lives – but there are solutions. From changing how we get our energy to limiting deforestation are some of the key solutions to climate change.

How will climate change affect the UK?

Those heatwaves, storms, flooding and long freezing winters the UK's getting more of now? These are all the effects of climate change.

What is the UK doing about climate change?

All countries need to reduce their greenhouse gas emissions that contribute to global warming. So how's the UK doing?

The difference between global warming and climate change

By Arna Lorenz

'Global warming' and 'climate change' are terms that we've become accustomed to hearing in recent years. In fact, a simple Google search returns over one billion combined results, proving that phrases that 30 years ago were virtually unknown have now become ingrained into the English language.

However, global warming and climate change are phrases that are often confused, misunderstood or even used interchangeably. In summary, global warming describes the increase in the surface temperature across the planet, predominantly the result of high levels of greenhouse gases in the atmosphere. This is only one aspect of climate change, which is the long-term changes in regional or global climate, especially rainfall, wind and temperature.

Since the two processes are linked, and one is even the result of the other, it's unsurprising that people often fail to understand the difference between climate change and global warming.

Global warming

The term global warming first became prominent in the media in the 1980s, although it was coined a decade or more earlier, mainly in response to growing scientific awareness of the damage pollutants – particularly chlorofluorocarbons (CFCs), commonly used in aerosols and refrigerants – were having on the Earth's ozone layer.

The Earth's surface heats during the daytime as it is struck by rays from the sun. At night, the energy from the sun is radiated back into space, theoretically allowing the surface of the planet to cool and the temperature to be maintained at an optimum level.

However, the presence of greenhouses gases in the atmosphere, including carbon dioxide, methane and chlorofluorocarbons, causes the heat radiated from the surface of the Earth to be radiated back, in effect creating a gaseous shield around the Earth that prevents the sun's heat from being able to escape.

The increase over decades of greenhouses gases in the atmosphere, combined with factors such as the orbit of the Earth and changes in the energy output of the sun, has contributed significantly to heat retention at the surface level of the planet, causing the temperature of Earth to rise.

According to the National Oceanic and Atmospheric Association, since 1880 the average surface temperature of the Earth has increased by approximately 0.95 degrees Celsius (1.71 degrees Fahrenheit). While such a small rise in the temperature may seem inconsequential, the effects of such a tiny shift can be dramatic, particularly on the planet's climate.

Climate change

The differences between climate change and global warming are, in part, commonly misunderstood because the increased temperature leads directly to a changing climate (often seen in extremes of weather). Whereas the weather is the particular conditions in the atmosphere at a specific point of time (for example, to describe a Monday morning as 'cloudy with a little light rain'), the climate describes the conditions in the atmosphere for an extended period of time, such as over 30 years.

Weather changes very quickly, sometimes within hours, whereas climate patterns tend to last for decades. Consequently, the climate changes more slowly and long-term trends must be considered to produce an accurate understanding of them.

Climates can also be regional or global. For climate change to be established, at least one of the climatic variables – rainfall, wind or temperature – would need to fluctuate over an extended period in the same place, whether this is in a region of the Earth or across the planet as a whole.

For example, a sustained increase in rainfall in a previously arid region of Australasia for several decades would be classified as climate change, even if the climate in other parts of the world remained stable.

Climate change sceptics often argue that the climate is constantly changing and, historically, the Earth has experienced extremes of weather many times before. In part, this is accurate: climate change occurs for several reasons, some of which are entirely natural and unpredictable such as the shifting of tectonic plates or volcanic activity.

Global warming causes many side effects in terms of our fluctuating climate. Melting glaciers at the poles, more frequent and violent tropical storms, above average temperatures during summer days in Europe and extended periods of drought in developing countries can all be attributed to the increase in the Earth's surface temperature.

While natural causes can sometimes be blamed for climate change, scientists have recently concluded that human activity is almost entirely responsible for global warming during the last 170 years.

To summarise the difference between climate change and global warming, global warming (the increase in the average temperature of the Earth's surface due to the presence of gases in the atmosphere) is a significant cause of climate change. And if global warming is the cause, climate change is the effect, the long-term fluctuation in weather patterns over decades, on a regional or a global scale.

The problem of global warming should not be underestimated as it is driving a potentially catastrophic change in the world's climates, putting the livelihoods and long-term existence of communities throughout across the planet in jeopardy.

7 April 2020

What are the effects of climate change?

The effects of climate change are with us right now. Millions are suffering already. And younger generations are being robbed of their future on a healthy, liveable planet.

The effects of climate change include extreme heat and drought, more rainfall and more frequent extreme weather events such as storms and floods.

Around the world, climate change already has (or will soon have) impacts on every aspect of human life. Wildfires and extreme temperatures are already impacting people's health. Drought and less fresh water mean it's harder to grow food.

Climate change is a consequence of rising average global temperatures caused by greenhouse gas emissions from burning fossil fuels. Small changes in temperatures, from only 1ºC above what they were before the industrial era, are already affecting the environment and people worldwide.

World leaders agreed at the 2015 Paris climate summit to limit temperatures to well below 2ºC, and 1.5ºC if at all possible. This is because there really is no safe level of warming. It's also because at 2ºC, a number of island nations in the Pacific Ocean are under threat of being swallowed up almost entirely by sea level rises. At Paris, the governments of these countries lobbied hard for a target of 1.5ºC to ensure their survival.

Hasn't the climate always changed?

Some people say the climate has always changed, and that's true. But what human activity is doing to the planet's atmosphere is different to anything that's happened before – which were mostly smaller, natural changes taking place over millions of years.

The average global temperature on Earth has already increased by a little over 1ºC since 1880, and most of that since 1975.

Because of this rapid climate change, wildfires are more likely to rage out of control, reducing forests to ash. The oceans are warming and the water is becoming more acidic, causing mass coral die-offs and the loss of breeding grounds for sea creatures.

Delicate ecosystems that are home to insects, plants and animals struggle to adapt quickly enough to the changing climate, putting one million species at risk of extinction.

All of this means that our food security, health and quality of life are all under threat.

What are the impacts of climate change on people?

The impacts of heatwaves, wildfires, droughts, storms, floods and sea-level rise is already devastating for many communities around the world.

Impacts are predicted to become catastrophic if governments cannot bring greenhouse gas emissions and global temperature rises under control.

Storms, floods and sea-level rise destroy homes and lives

In the UK, climate change is causing more extreme weather such as heatwaves and storms. Increased heat in the atmosphere leads to more heavy rainfall and more frequent flooding, often impacting the same areas over and over again.

Flooding has turned lives upside-down in Yorkshire, Somerset and Cumbria. And a coastal village in Wales, Fairbourne, is being evacuated because of the growing threat of sea level rise.

In the Pacific island nations, Central America, the Caribbean and southern US states such as Louisiana and Florida, hurricanes are increasingly severe and far more frequent, leading to loss of life, homes, harvests or entire farms.

In South and Southeast Asia, countries like Bangladesh, Myanmar and the Philippines face more and stronger cyclones or typhoons – which can destroy entire regions. It takes years to rebuild.

Extreme weather events are devastating to any community. But those in poorer nations, and people living in underprivileged communities in rich nations, also struggle to recover. In the US, government neglect, inequality and racism have led to inadequate emergency responses and very slow recovery. This was seen most starkly after Hurricane Katrina in 2010 and Hurricane Maria in 2017.

Sea level rises are already affecting Pacific island nations like Kiribati and the Marshall Islands.

Native Americans in Alaska and Louisiana are facing not only a rapid reshaping of their ancestral coastlines, but also the effects of polluting oil and gas drilling.

Indigenous Peoples worldwide suffer disproportionately from both the causes and effects of climate change. This is despite often having a deep understanding of how to look after nature and use natural resources sustainably.

Heat, drought and fires increase health risks

Heatwaves, droughts and wildfires have serious effects on people and communities. As well as the damage caused by fire itself to wildlife and human settlements, smoke inhalation and air pollution from fire is a major health risk.

Heat and drought are also bad for human health. There is a limit to how much heat a human body can cope with. And lack of clean water to drink or grow food can cause illness, malnourishment, famine, migration and war.

Extreme heat, particularly in cities, can be deadly – particularly for older people. In Europe summer heatwaves have already caused tens of thousands of deaths in some recent years.

Heatwaves also worsen droughts around the world. These extended periods without water threaten not only human health, but also how much food can be grown. According to the UN, this has led to alarming rises in global hunger.

Heatwaves and drought can lead to wildfires. Wildfires are already affecting many countries around the world on a regular basis. The Amazon, Australia, the Western US and Siberia have all seen alarming wildfires in recent years.

What are the effects of climate change on nature?

Extreme weather

A key effect of climate change is extreme weather.

Rising temperatures cause heatwaves droughts, and wildfires. They also warm the atmosphere, increasing moisture – which means more rainfall, storms and flooding.

Storms and flooding are affecting many parts of the world, including the UK. Even extreme cold weather is also thought to be another effect of climate change.

Fire is particularly merciless when it tears through any landscape, killing or harming thousands of species of plants and those animals unable to escape. It is estimated that the Australia fires of early 2020 killed or harmed nearly three billion animals. Some of the fires were so enormous they even created their own weather events.

Polar and glacier ice melt and sea-level rise

Higher average global temperatures are also melting ice at the polar regions and in glaciers in mountainous regions.

In the Arctic, which now experiences heatwaves, the sea ice disappears almost entirely in summer – and the region is predicted to be completely ice-free by the mid-2030s.

Antarctic ice shelves have lost nearly 4 trillion metric tons of ice since the mid-1990s, with warming ocean waters melting them faster than they can refreeze.

Melting polar ice leads to sea-level rise around the world, which (along with increased storm surges) is starting to permanently re-shape coastal regions. This is already happening in the Arctic, the South Pacific and parts of the southern US.

The above information is reprinted with kind permission from Greenpeace.
© 2023 Greenpeace

www.greenpeace.org.uk

Climate change is already altering everything, from fertility choices to insuring our homes

By Stefan Ellerbeck

- **Climate change is already affecting people's lives in a variety of ways.**

- **Global warming is the biggest health threat facing humanity, the World Health Organization says.**

- **It's also making people rethink family planning choices and putting properties at risk of becoming uninsurable.**

- **Disruptions to supply chains because of extreme weather are shaking the global economy.**

How is climate change affecting you?

You may think the biggest impacts lie far away – in terms of time or geography. But global warming is already changing the way many of us live or think.

1. Health suffers because of climate change

Climate change is the biggest health threat facing humanity, the World Health Organization says, estimating that it will cause around a quarter of a million additional deaths each year in 2030-50. These will mainly be from malnutrition, malaria, diarrhoea and heat stress.

However, climate change is already having more subtle effects on health and wellbeing. Spring is beginning earlier in many places, meaning there's a higher pollen count. This is bad news for allergy sufferers. Higher temperatures in the United States made the pollen season 11-27 days longer between 1995 and 2011, the Asthma and Allergy Foundation of America says.

Rising temperatures also contribute to worsening air quality, which can increase the risk and severity of asthma attacks.

2. Climate change is raising the cost of living

COVID-19 has received most of the blame for recent global supply chain problems, but climate change is also having an impact. When supply chains are shaken, this impacts the availability and cost of goods.

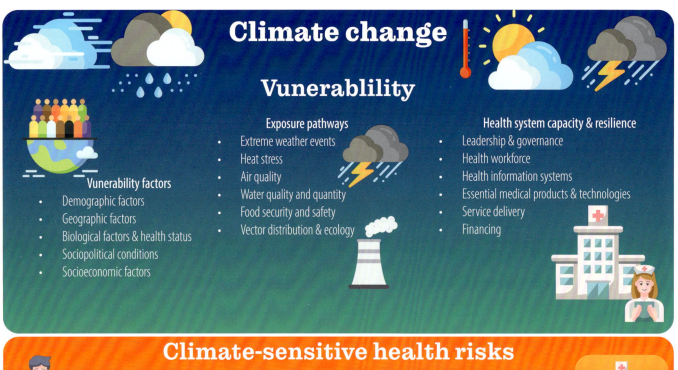

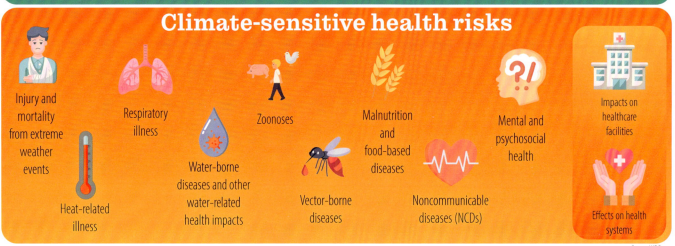

Source: WE Forum

Do climate change-related factors impact fertility decisions?

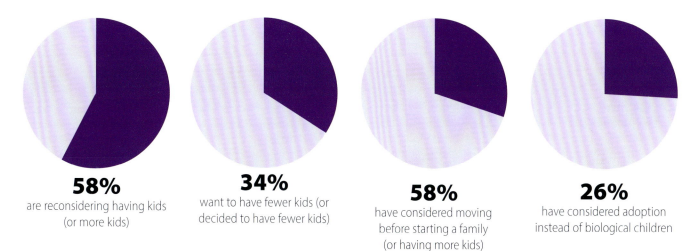

58%
are reconsidering having kids
(or more kids)

34%
want to have fewer kids (or
decided to have fewer kids)

58%
have considered moving
before starting a family
(or having more kids)

26%
have considered adoption
instead of biological children

Source: A survey of 2,800+ women in the US, by Modern Fertility

Freezing weather in Texas in February 2021 triggered the United States' most severe energy blackout of all time, leading to shutdowns at three major semiconductor plants and adding to the global shortage of microchips.

The cost of living is also soaring because of the global surge in energy prices. While Russia's war on Ukraine is driving much of this now, climate change is also a factor.

'Companies face up to $120 billion in costs from environmental risks in their supply chains by 2026,' according to research published in 2021 by CDP, a nonprofit that runs the world's largest environmental disclosure system. This will include increased costs for raw materials, and because of regulatory changes such as carbon pricing as the world addresses environmental crises, the report says.

In 2021, more than 20% of American adults lived in households unable to pay their utility bills.

3. Warming oceans are threatening our way of life

Sea level rises could pose the biggest threat to global supply chains, potentially putting ports and coastal infrastructure out of action. Higher sea temperatures may also cause more severe storms in tropical parts of the world, posing a threat to life and infrastructure.

The sea is home to most of our biodiversity, and 3 billion people globally rely on it for their livelihoods, according to the UN. However, carbon emissions from human activity are causing ocean warming, acidification and oxygen loss, putting large numbers of marine-related jobs at risk, it says.

4. People might have fewer babies

People are increasingly citing the climate crisis as a major reason why they may decide to have fewer or even no children. According to a study in the United States, a third of women said they will reduce their anticipated family size because of it.

However a similar number of the more than 2,800 American women surveyed by Modernfertility.com said the issue has made them decide to have children sooner. The study says this is because it's either made them focus more on what's important to them or given them a sense of urgency.

5. Your property could become uninsurable

Insurance is something that nearly everyone has, but climate change poses a 'systemic risk' to the sector, according to professional services company Grant Thornton.

Extreme weather events led to insured losses of $105 billion in 2021, the fourth-highest level since 1970, according to preliminary estimates by Swiss Re, one of the world's leading providers of reinsurance and insurance.

This not only potentially makes insurance more expensive for everyone, but it also means some assets could become uninsurable. One in 25 Australian homes could be uninsurable by 2030, according to the Climate Council.

6. Increased chance of another pandemic

Climate change makes new pandemics more likely, because as temperatures increase, wild animals will be forced to change habitats. This could lead to them living nearer to human populations, increasing the chances of a virus jumping between species and causing the next pandemic, according to a report published by the scientific journal Nature.

'Geographic range shifts' will mean mammals encounter each other for the first time, and in doing so will share thousands of viruses, the report says. Even keeping global warming under 2°C this century 'will not reduce future viral sharing', the scientists note.

The World Economic Forum is committed to helping limit global warming to 1.5°C above pre-industrial levels to stave off catastrophe. It aims to work with leaders to increase climate commitments, collaborate with partners to develop private initiatives, and provide a platform for innovators to realize their ambition and contribute solutions.

24 June 2022

UK Heatwave: Britain's 40C future and why our homes, trains and schools aren't prepared

From moving the date of school exams to painting railway tracks white and retrofitting hospitals, the UK has its work cut out preparing for extreme heat.

By Madeleine Cuff, Environment Correspondent

Temperatures next week could hit 40°C in parts of the UK in what forecasters are warning is a 'red alert' heatwave that will bring serious disruption to everyday life.

Such temperatures are making headlines as an unprecedented extreme right now, but in a few years' time they could become a 'new normal' for parts of the country.

'By 2050, we will regularly have temperatures above 35°C in the south of the UK,' predicts Chloe Brimicombe, a heat stress researcher at the University of Reading.

Heat-related deaths, which currently number about 2,000 a year in the UK, will soar, she tells i: 'By the 2050s between 5,000 and 7,000 people will die annually from extreme heat every year.'

Britons may talk a lot about the weather, but the UK is not a country designed to withstand such extremes. Homes are built to stay warm during relatively mild winters, while infrastructure such as trains and power lines can struggle in hot weather.

Scientists have warned for years that the UK is not prepared to cope with the fierce summer heat climate change is driving. From overheating homes and buckled rail lines to power cuts, mobile service outage, and sweltering hospitals, many services currently are pushed to the point of crisis in high temperatures.

Major overhauls are needed across UK infrastructure to prevent unnecessary deaths and huge economic impacts in the coming decades, scientists stress.

'Heat scientists have for years been telling the government to prepare for this. This event is happening, we have very little preparation in place,' says Brimicombe. 'We need mass mobilisation and real leadership from whoever the next government is – they need to listen to the advice of scientists urgently.'

Maximum summer temperatures
UK, business-as-usual climate scenario, 2010-2099

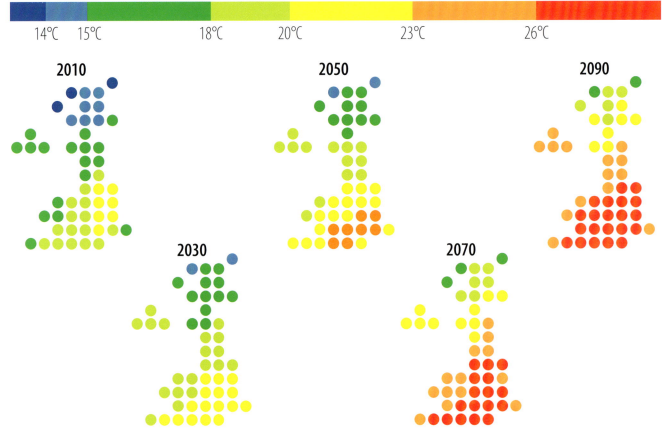

Source: Met Office/Hadley Centre Model

Housing

In the past decade, millions of new homes were built that are not designed to cope with rising temperatures, which has condemned families to living in unbearably hot homes during the summer months.

In December, the Government finally took action, changing building regulations so developers must now consider the risk of overheating when designing new build properties.

But existing homes are still a problem, says Professor Kevin Lomas from the University of Loughborough. He conducted research last year that suggested almost 20 per cent of bedrooms in English homes overheat in the summer months.

'We live in a cool damp country, and so our housing stock is built primarily for trying to keep us warm in winter, with almost no consideration of hot summer conditions,' he tells i.

He says efforts to retrofit homes with insulation and heat pumps will not only cut carbon and keep homes warmer in the winter months, but will also help to manage summertime temperatures too. Heat pumps, after all, can cool homes as well as heat them.

But Professor Lomas argues more government action is needed to make this a reality, echoing repeated calls from the government's climate advisors for ministers to take action on home retrofit.

'If you are shading and ventilating, then insulation works both ways,' Professor Lomas says. 'The Government should be more keen than it is … to retrofit houses so that they are more energy-efficient.'

Utilities

Hot weather puts extreme strain on key utilities such as power and water. Water companies must manage households filling paddling pools and taking multiple showers a day without letting their reservoirs run dry. Meanwhile, for every degree increase in temperature, UK power demand jumps 0.9 per cent as people switch on air conditioners and fans.

To complicate matters further, extreme heat can cause interruptions to power supplies as overhead power cables sag and key electronic equipment fails. If the power system fails, that causes a 'cascade failure' with shocks felt cross the economy, the Climate Change Committee warns.

Persuading households to use water and power wisely during periods of extreme weather will be crucial for securing supplies in the future, experts say. Meanwhile a programme of upgrades is already under way across the electricity system to make it more resilient to climate effects.

Healthcare

Hospitals have cancelled surgeries planned for next week because of the hot weather, with hospital bosses warning it is a 'challenge' to keep wards and departments cool for elderly citizens.

They are right to be cautious, as evidence shows Britain's hospitals and care settings are prone to overheating. Between 2019 and 2020 there were 3,600 incidences of hospitals overheating, defined as inside temperatures above 26°C. Meanwhile, nine in 10 hospital wards are at risk of overheating under current climate conditions.

More effort is needed to retrofit care settings with external shading such as trees, shutters and awnings, and to improve ventilation, experts say. The good news is early research suggests hospitals built in the 1960s can be successfully retrofitted with better insulation, ventilation and shading to cope with increasingly hot summers over the next 20 years.

Transport

Most of the UK's rail network can operate when track temperatures heat up to 46°C, roughly equivalent to air temperature of around 30°C. In the summer of 2019, when temperatures hit a record 38.7°C in the UK, rail lines buckled in the heat causing widespread disruption across the rail network.

Train companies are commonly forced to introduced speed restrictions during heatwave conditions, causing knock-on delays and complications.

Network Rail, which manages the railway infrastructure, warns there is a 'major' risk of trains derailing and railway verges catching fire in the coming decades as climate change advances.

To counter the problem, network operators are painting rails white to reflect the heat, installing tracks on reinforced concrete slabs to prevent buckling, and upgrading signalling systems.

Education

A 2021 UK study found that most classrooms – even new-build classrooms – experienced overheating for more than 40 per cent of school hours.

The Government has changed design standards to force new and refurbished schools to be designed with higher temperatures in mind, but teachers say more work is needed to ensure all classrooms are kept cool and well ventilated in the summer months.

That might mean shifting the school year around, for example, to hold exams in the cooler months, according to Richard Miller, head of adaptation at the Climate Change Committee. 'These [heatwave] events often coincide with exam times, which is one of the key determiners of future life chances and earnings for young people,' he told i.

Moving critical exams to a cooler time of the year is 'definitely the kind of thing we should be considering', he added.

15 July 2022

Brainstorm

In pairs, think of things that your school could do to prepare for extreme temperatures. Consider both summer and winter and create a brainstorm of your ideas.

What is causing our climate to change so quickly now?

By Rita Nogherotto, Clara Burgard & Chris D. Jones

Abstract

Our planet's climate has been warming much faster over the past 100 years than it has over the 10,000 years before that. In this article, we will explore how we know that the climate has changed so quickly in the last century, what carbon dioxide (CO2) has to do with climate change, and why humans are responsible for the recent increase of CO2 in the atmosphere. Understanding the problem is the best way to find a solution!

Global Warming or Climate Change?

It is about 1°C warmer today on Earth than 60 years ago. 1°C can seem like a very small change. After all, the difference in temperature between winter and summer can be 30°C in some regions! So why does global warming of a few degrees worry scientists so much?

Do you remember the last time you had fever? The usual human body temperature is around 37°C. When you have a fever, your body gets warmer by 1 or 2°C but you cannot concentrate well anymore, your body does not work as usual, and you need to rest. In the case of your body, it is a matter of a few degrees between being healthy and being sick. The effect of global warming on the Earth is like that of a fever: when the air at the surface gets warmer, the entire planet does not work as usual. The oceans warm as well. The thick ice melts away in cold and mountainous regions. All over the planet, living beings need to get used to new conditions or move to other regions. The warming of the planet affects everything within it, which is why we prefer to speak of climate change rather than of global warming. It is not only a problem of air temperature – there are changes in all elements of the environment around us.

Climate change at present

In the past, the Earth has been warmer and colder than it is today. These temperature differences were due to changes in the Earth's position compared to the Sun, or to large natural events such as volcanic eruptions. These periods of warming and cooling occurred over several thousands of years. Nature and humans could therefore slowly adapt to the changing climate conditions.

This time it is different: climate change is happening very quickly. In Figure 1, you can see a reconstruction of the Earth's average surface temperature over the past 11,400 years. You can easily see that the temperature has risen much faster

Variations in the Earth's average surface temperature over the past 11,400 years

Figure 1

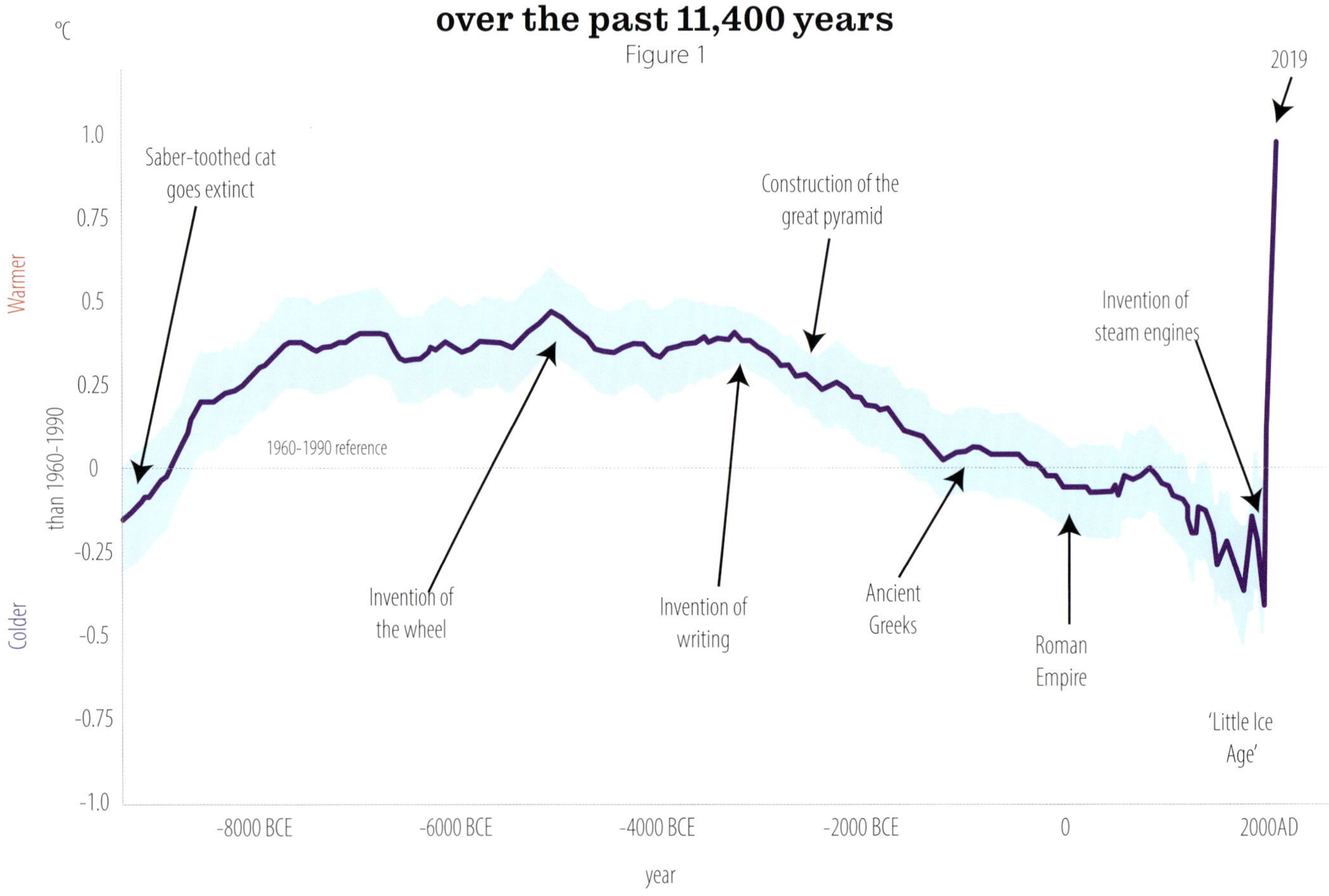

Note: The purple line is the mean and the shading around it is the uncertainty around the mean. BCE means before common era; AD means Anno Domini (Data until 1,900 compiled by [1]; data for 1900 to 2019 from [2]).
Source: Frontiers

Carbon dioxide concentration at Mauna Loa observatory

Figure 2

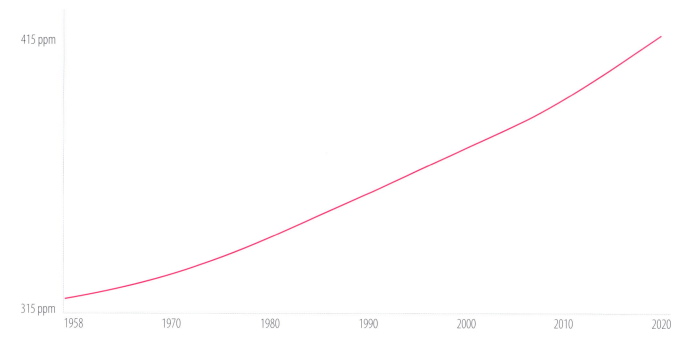

Note: Figure 2 - The atmospheric CO2 concentration, as measured on Mauna Loa, has been rising since it was first measured in 1958 (Data from [3]).
Source: Frontiers

in the last decades compared to the temperatures that the Earth experienced previously. It only took 60 years to warm the Earth by around 1°C. Scientists have calculated that the Earth might be 3°C warmer–or even more–in 80 years, if we continue living like we are at the moment. Keep in mind that ice ages were only about 4°C cooler than today's temperatures, so this is actually a big deal! But…what is the difference between today's climate change and past climate changes? And what do humans have to do with it?

To answer these questions, let us go back some years in time. In the 1950s, a young scientist named Charles Keeling was wondering why the Earth had been warming so much over the past 100 years. He decided to measure the amount of carbon dioxide (CO2) in the atmosphere, or what scientists call the atmospheric CO2 concentration. Why CO2? Charles knew that this gas is a greenhouse gas, meaning that it has a role in adjusting the temperature of Earth's atmosphere. The more CO2 in the atmosphere, the more heat the atmosphere can absorb and send back to the Earth's surface, leading to global warming.

From that moment on, Charles measured the atmospheric CO2 concentration every day–for years. Every year, he saw more atmospheric CO2 than in the year before (Figure 2). This increase continues today: in 2019, atmospheric CO2 concentrations were higher than they were at any time in at least 2 million years [4].

Why is there more and more CO2 in the atmosphere?

Charles discussed his findings with other scientists. They knew that more CO2 in the atmosphere meant that the Earth was warming. And they knew that such warming could have dramatic consequences on the whole climate. They decided to find out what was causing this increase in CO2. They hoped that tracking down the cause might help prevent more CO2 from getting into the atmosphere.

As a first step, the scientists looked at the amounts of CO2 in the atmosphere in the past. They found that, at least for the last 10,000 years (and probably for the last million years!), the amount of CO2 in the atmosphere had never been higher than the amount Charles measured!

In the past, CO2 was added and removed from the atmosphere through natural processes. Plants and oceans, for example, can store CO2 and regularly exchange it with the atmosphere. Sometimes plants and oceans take up CO2 from the atmosphere, reducing atmospheric CO2 concentration, and sometimes plants and oceans release CO2, increasing atmospheric CO2 again.

These natural processes are very slow–they can take thousands of years to absorb or release large amounts of CO2. This is why natural processes cannot possibly explain the rapid accumulation of CO2 in the atmosphere during the last century. If the CO2 is not coming from nature, then where is it coming from?

The CO2 concentration started to rise very quickly around 150 years ago, when we started using machines to accomplish more work in a shorter time. To run these machines, we needed energy. We got this energy from burning oil, coal, and natural gas, which are called fossil fuels. When we burn fossil fuels, CO2 and other greenhouse gases, including methane and nitrogen dioxide, are released (emitted) into the atmosphere.

Over time, we needed more energy to power our cars, to heat our homes, to charge our smartphones, and so on. So, we burned more fossil fuels, leading to more greenhouse gas emissions. In parallel with industrial development, the human population has grown rapidly. To feed more and more humans, forests had to be cut down to create large fields for agriculture. As a result, less CO2 could be taken up by trees. It is therefore we humans who were behind the fast increase in atmospheric CO2 over the last century.

And, since we have not yet stopped burning fossil fuels, we are still responsible for adding more and more CO2 into the atmosphere!

The amount of CO2 we emit is very small compared to that exchanged naturally between plants, oceans, and the atmosphere. So why does this small amount emitted by humans have such a big impact on our planet's temperature? Cannot plants and oceans just take up more CO2? Let us take a sneak peek into the natural 'CO2-control task force' and how the atmosphere, plants, and the oceans divide the work between them.

Humans introduced an imbalance in natural CO2 exchange

Nature likes a balance between different elements. For example, when you open your window in winter, cold air from outside and warm air from inside mix until they reach a common temperature–the indoor and outdoor temperatures are balanced. Over thousands of years, nature adapted to keep such a balance in the Earth's CO2.

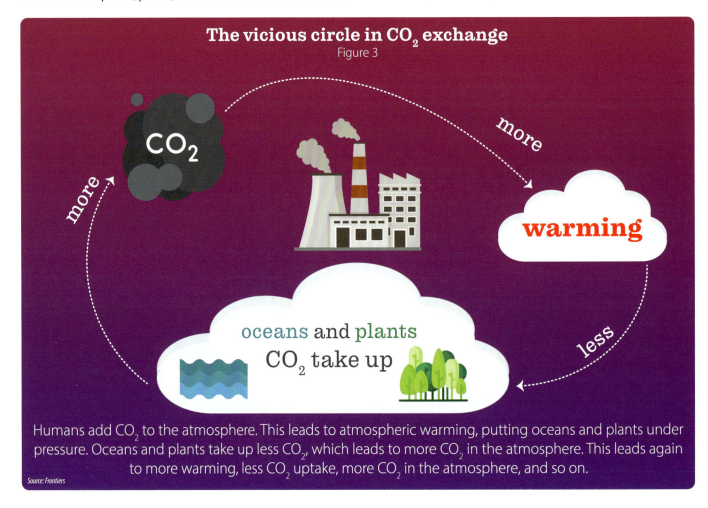

The vicious circle in CO_2 exchange
Figure 3

Humans add CO_2 to the atmosphere. This leads to atmospheric warming, putting oceans and plants under pressure. Oceans and plants take up less CO_2, which leads to more CO_2 in the atmosphere. This leads again to more warming, less CO_2 uptake, more CO_2 in the atmosphere, and so on.

Source: Frontiers

Glossary

Global Warming: The warming we are experiencing due to the increased greenhouse effect.

Climate Change: The change of the Earth's climate due to the increased greenhouse effect causing warming of the Earth's surface.

Carbon Dioxide: A gas in the atmosphere that is created when we burn fossil fuels. CO_2 is a greenhouse gas and can stay in the air for many years. It is the main cause of climate change.

Atmospheric CO_2 Concentration: When measuring gases like carbon dioxide, the term concentration is used to describe the amount of gas in a given volume of air. It is usually measured in ppm (part per million), which is a way of expressing very dilute concentrations of substances. Just as per cent means out of a hundred, parts per million means out of a million.

Greenhouse Gas: A gas in the atmosphere which can absorb heat and cause the planet to warm up. These occur naturally, such as carbon dioxide and water vapor, but human activity is putting more greenhouse gases into the air leading to the planet getting warmer.

Fossil Fuel: Fuels, such as coal, oil, and natural gas that were formed millions of years ago when plants and animals died and became buried. Burning fossil fuels creates CO_2, which goes into the atmosphere.

Vicious Circle: A chain of events in which a change compared to the usual situation has the effect of creating new problems which then cause the original change to change more, creating new problems and more change again and again.

Carbon Cycle: The movement of carbon through nature–plants absorb CO_2 from the air as they grow, and release it again when they die. The water in the oceans dissolves CO_2. This cycle is called the carbon cycle.

If the amount of CO_2 in the atmosphere was higher than usual, the plants and the oceans would take up more CO_2 than usual, until CO_2 amounts were all in balance again. The atmosphere, plants, and the oceans form what could be considered a natural CO_2-control task force, because together they regulate the exchange of CO_2.

When humans started burning fossil fuels, more CO_2 started accumulating in the atmosphere in a short period of time. The change in atmospheric CO_2 was so fast that the CO_2-control task force could not keep up with establishing the balance. The fast rise in atmospheric CO_2 concentration and the resulting atmospheric warming put the plants and the oceans under pressure. The task force did not have time to adapt to higher amounts of CO_2, as it could in the past. As a result, the atmospheric CO_2 concentration increased even more, leading to more atmospheric warming, putting plants and oceans under even more pressure, and so on. This is what we call the vicious circle of the carbon cycle (Figure 3). This vicious circle has only a small effect compared to the human

CO_2 emissions coming from the burning of fossil fuels, but it is expected to have a larger effect in the future! You should know that there are many such vicious circles in nature. For example, different vicious circles are responsible for the very fast warming in polar regions. Other greenhouse gases are also increased by human activity, although CO_2 is the biggest cause of climate change. Methane (CH_4) and nitrous oxide (N_2O) have also increased a lot since pre-industrial times because of the burning of fossil fuels and the increase in agriculture.

In conclusion

The amount of CO_2 that humans send into the atmosphere may appear small at first glance. However, it is enough to throw off the balance between the atmosphere, the oceans, and plants.

The good news is that Charles' measurements started a scientific investigation into the changes in atmospheric CO_2. Ever since Charles' first measurements in the 1950s, the crowd of scientists trying to understand climate change has grown. Today, we know much more than we knew in Charles' time. We know that it is humans who are behind the rise in atmospheric CO_2 concentration. And it is therefore also humans who will be able to stop climate change. By reducing CO_2 emissions, we can slow down climate change: every ton of CO_2 that does not go into the atmosphere means less warming. Our actions now are essential to avoid future climate change [5]! We need smart, creative, and hopeful people to think of new ways to live our lives! Maybe you will be one of those people?

Brief Summary

Climate change: we hear about it almost every day now. We hear that our planet is warming and that this warming is caused by us. But why is that a problem? And what exactly are the mechanisms driving climate change at the moment? In this article we try to explain to a younger audience (8–11 years old) what are the causes of the recently observed changes in our climate. These changes are occurring faster than past changes. They cannot be caused by natural events only. Understanding the causes of climate change is crucial to figure out what is going on and what can be done to limit its consequences.

21 February 2022

www.kids.frontiersin.org

How earth's climate changes naturally (and why things are different now)

Earth's climate has fluctuated through deep time, pushed by these 10 different causes. Here's how each compares with modern climate change.

By Howard Lee

Earth has been a snowball and a hothouse at different times in its past. So if the climate changed before humans, how can we be sure we're responsible for the dramatic warming that's happening today?

In part it's because we can clearly show the causal link between carbon dioxide emissions from human activity and the 1.28 degree Celsius (and rising) global temperature increase since preindustrial times. Carbon dioxide molecules absorb infrared radiation, so with more of them in the atmosphere, they trap more of the heat radiating off the planet's surface below.

But paleoclimatologists have also made great strides in understanding the processes that drove climate change in Earth's past. Here's a primer on 10 ways climate varies naturally, and how each compares with what's happening now.

Solar Cycles

Magnitude: 0.1 to 0.3 degrees Celsius of cooling

Time frame: 30- to 160-year downturns in solar activity separated by centuries

Every 11 years, the sun's magnetic field flips, driving an 11-year cycle of solar brightening and dimming. But the variation is small and has a negligible impact on Earth's climate.

More significant are 'grand solar minima,' decades-long periods of reduced solar activity that have occurred 25 times in the last 11,000 years. A recent example, the Maunder minimum, which occurred between 1645 and 1715, saw solar energy drop by 0.04% to 0.08% below the modern average. Scientists long thought the Maunder minimum might have caused the 'Little Ice Age,' a cool period from the 15th to the 19th century; they've since shown it was too small and occurred at the wrong time to explain the cooling, which probably had more to do with volcanic activity.

The sun has been dimming slightly for the last half-century while the Earth heats up, so global warming cannot be blamed on the sun.

Volcanic Sulfur

Magnitude: Approximately 0.6 to 2 degrees Celsius of cooling

Time frame: 1 to 20 years

In the year 539 or 540 A.D., the Ilopango volcano in El Salvador exploded so violently that its eruption plume reached high into the stratosphere. Cold summers, drought, famine and plague devastated societies around the world.

Eruptions like Ilopango's inject the stratosphere with reflective droplets of sulfuric acid that screen sunlight, cooling the climate. Sea ice can increase as a result, reflecting more sunlight back to space and thereby amplifying and prolonging the global cooling.

Ilopango triggered a roughly 2 degree Celsius drop that lasted 20 years. More recently, the eruption of Pinatubo in the Philippines in 1991 cooled the global climate by 0.6 degrees Celsius for 15 months.

Volcanic sulfur in the stratosphere can be disruptive, but in the grand scale of Earth's history it's tiny and temporary.

Short-Term Climate Fluctuations

Magnitude: Up to 0.15 degrees Celsius

Time frame: 2 to 7 years

On top of seasonal weather patterns, there are other short-term cycles that affect rainfall and temperature. The most significant, the El Niño–Southern Oscillation, involves circulation changes in the tropical Pacific Ocean on a time frame of two to seven years that strongly influence rainfall in North America. The North Atlantic Oscillation and the Indian Ocean Dipole also produce strong regional effects. Both of these interact with the El Niño–Southern Oscillation.

The interconnections between these cycles used to make it hard to show that human-caused climate change was statistically significant and not just another lurch of natural variability. But anthropogenic climate change has since gone well beyond natural variability in weather and seasonal temperatures. The U.S. National Climate Assessment in 2017 concluded that there's 'no convincing evidence for natural cycles in the observational record that could explain the observed changes in climate.'

Orbital Wobbles

Magnitude: Approximately 6 degrees Celsius in the last 100,000-year cycle; varies through geological time

Time frame: Regular, overlapping cycles of 23,000, 41,000, 100,000, 405,000 and 2,400,000 years

Earth's orbit wobbles as the sun, the moon and other planets change their relative positions. These cyclical wobbles, called Milankovitch cycles, cause the amount of sunlight to vary at middle latitudes by up to 25% and cause the climate to oscillate. These cycles have operated throughout time, yielding the alternating layers of sediment you see in cliffs and road cuts.

During the Pleistocene epoch, which ended about 11,700 years ago, Milankovitch cycles sent the planet in and out of ice ages. When Earth's orbit made northern summers warmer than average, vast ice sheets across North America, Europe and Asia melted; when the orbit cooled northern

summers, those ice sheets grew again. Since warmer oceans dissolve less carbon dioxide, atmospheric carbon dioxide levels rose and fell in concert with these orbital wobbles, amplifying their effects.

Today Earth is approaching another minimum of northern sunlight, so without human carbon dioxide emissions we would be heading into another ice age within the next 1,500 years or so.

Faint Young Sun

Magnitude: No net temperature effect

Time frame: Constant

Though the sun's brightness fluctuates on shorter timescales, it brightens overall by 0.009% per million years, and it has brightened by 48% since the birth of the solar system 4.5 billion years ago.

Scientists reason that the faintness of the young sun should have meant that Earth remained frozen solid for the first half of its existence. But, paradoxically, geologists have found 3.4-billion-year-old rocks that formed in wave-agitated water. Earth's unexpectedly warm early climate is probably explained by some combination of less land erosion, clearer skies, a shorter day and a peculiar atmospheric composition before Earth had an oxygen-rich atmosphere.

Clement conditions in the second half of Earth's existence, despite a brightening sun, do not create a paradox: Earth's weathering thermostat counteracts the effects of the extra sunlight, stabilizing Earth's temperature (see next section).

Carbon Dioxide and the Weathering Thermostat

Magnitude: Counteracts other changes

Time frame: 100,000 years or longer

The main control knob for Earth's climate through deep time has been the level of carbon dioxide in the atmosphere, since carbon dioxide is a long-lasting greenhouse gas that blocks heat that tries to rise off the planet.

Volcanoes, metamorphic rocks and the oxidization of carbon in eroded sediments all emit carbon dioxide into the sky, while chemical reactions with silicate minerals remove carbon dioxide and bury it as limestone. The balance between these processes works as a thermostat, because when the climate warms, chemical reactions become more efficient at removing carbon dioxide, putting a brake on the warming. When the climate cools, reactions become less efficient, easing the cooling. Consequently, over the very long term, Earth's climate has remained relatively stable, providing a habitable environment. In particular, average carbon dioxide levels have declined steadily in response to solar brightening.

However, the weathering thermostat takes hundreds of thousands of years to react to changes in atmospheric carbon dioxide. Earth's oceans can act somewhat faster to absorb and remove excess carbon, but even that takes millennia and can be overwhelmed, leading to ocean acidification. Each year, the burning of fossil fuels emits about 100 times more carbon dioxide than volcanoes emit – too much too fast for oceans and weathering to neutralize it, which is why our climate is warming and our oceans are acidifying.

Plate Tectonics

Magnitude: Roughly 30 degrees Celsius over the past 500 million years

Time frame: Millions of years

The rearrangement of land masses on Earth's crust can slowly shift the weathering thermostat to a new setting.

The planet has generally been cooling for the last 50 million years or so, as plate tectonic collisions thrust up chemically reactive rock like basalt and volcanic ash in the warm, wet tropics, increasing the rate of reactions that draw carbon

The cooling current

Long ago, South America and Australia were right next to Antartica, as seen in the light blue contours below. The separation of these continents allowed the Antarctic Circumpolar Current to develop, which enchanced ocean circulation and led to global cooling.

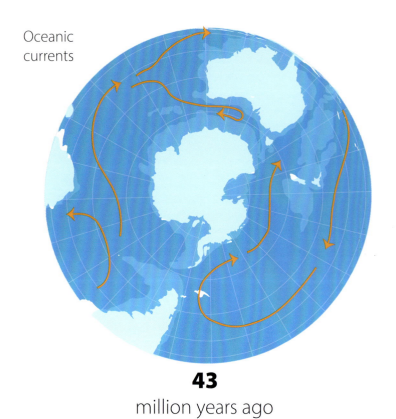

Oceanic currents

43
million years ago

Antarctic Circumpolar Current

Tasman Gateway

25
million years ago

Source: Quanta Magazine

dioxide from the sky. Additionally, over the last 20 million years, the building of the Himalayas, Andes, Alps and other mountains has more than doubled erosion rates, boosting weathering. Another contributor to the cooling trend was the drifting apart of South America and Tasmania from Antarctica 35.7 million years ago, which initiated a new ocean current around Antarctica. This invigorated ocean circulation and carbon dioxide-consuming plankton; Antarctica's ice sheets subsequently grew substantially.

Earlier, in the Jurassic and Cretaceous periods, dinosaurs roamed Antarctica because enhanced volcanic activity, in the absence of those mountain chains, sustained carbon dioxide levels around 1,000 parts per million, compared to 415 ppm today. The average temperature of this ice-free world was 5 to 9 degrees Celsius warmer than now, and sea levels were around 250 feet higher.

Asteroid Impacts

Magnitude: Approximately 20 degrees Celsius of cooling followed by 5 degrees Celsius of warming (Chicxulub)

Time frame: Centuries of cooling, 100,000 years of warming (Chicxulub)

The Earth Impact Database recognizes 190 craters with confirmed impact on Earth so far. None had any discernable effect on Earth's climate except for the Chicxulub impact, which vaporized part of Mexico 66 million years ago, killing off the dinosaurs. Computer modeling suggests that Chicxulub blasted enough dust and sulfur into the upper atmosphere to dim sunlight and cool Earth by more than 20 degrees Celsius, while also acidifying the oceans. The planet took centuries to return to its pre-impact temperature, only to warm by a further 5 degrees Celsius, due to carbon dioxide in the atmosphere from vaporized Mexican limestone.

How or whether volcanic activity in India around the same time as the impact exacerbated the climate change and mass extinction remains controversial.

Evolutionary Changes

Magnitude: Depends on event; about 5 degrees Celsius cooling in late Ordovician (445 million years ago)

Time frame: Millions of years

Occasionally, the evolution of new kinds of life has reset Earth's thermostat. Photosynthetic cyanobacteria that arose some 3 billion years ago, for instance, began terraforming the planet by emitting oxygen. As they proliferated, oxygen eventually rose in the

atmosphere 2.4 billion years ago, while methane and carbon dioxide levels plummeted. This plunged Earth into a series of 'snowball' climates for 200 million years. The evolution of ocean life larger than microbes initiated another series of snowball climates 717 million years ago – in this case, it was because the organisms began raining detritus into the deep ocean, exporting carbon from the atmosphere into the abyss and ultimately burying it.

When the earliest land plants evolved about 230 million years later in the Ordovician period, they began forming the terrestrial biosphere, burying carbon on continents and extracting land nutrients that washed into the oceans, boosting life there, too. These changes probably triggered the ice age that began about 445 million years ago. Later, in the Devonian period, the evolution of trees further reduced carbon dioxide and temperatures, conspiring with mountain building to usher in the Paleozoic ice age.

Large Igneous Provinces

Magnitude: Around 3 to 9 degrees Celsius of warming

Time frame: Hundreds of thousands of years

Continent-scale floods of lava and underground magma called large igneous provinces have ushered in many of Earth's mass extinctions. These igneous events unleashed an arsenal of killers (including acid rain, acid fog, mercury poisoning and destruction of the ozone layer), while also warming the planet by dumping huge quantities of methane and carbon dioxide into the atmosphere more quickly than the weathering thermostat could handle.

In the end-Permian event 252 million years ago, which wiped out 81% of marine species, underground magma ignited Siberian coal, drove up atmospheric carbon dioxide to 8,000 parts per million and raised the temperature by between 5 and 9 degrees Celsius. The more minor Paleocene-Eocene Thermal Maximum event 56 million years ago cooked methane in North Atlantic oil deposits and funneled it into the sky, warming the planet by 5 degrees Celsius and

acidifying the ocean; alligators and palms subsequently thrived on Arctic shores. Similar releases of fossil carbon deposits happened in the end-Triassic and the early Jurassic; global warming, ocean dead zones and ocean acidification resulted.

If any of that sounds familiar, it's because human activity is causing the same effects today.

As a team of researchers studying the end-Triassic event wrote in April in Nature Communications, 'Our estimates suggest that the amount of CO_2 that each … magmatic pulse injected into the end-Triassic atmosphere is comparable to the amount of anthropogenic emissions projected for the 21st century.'

21 July 2020

Original story reprinted with permission from *Quanta Magazine*, an editorially independent publication supported by the Simons Foundation.

© 2023 Quanta Magazine

www.quantamagazine.org

What Can Be Done?

Climate change: World has the money and the means to stop crisis but isn't doing so, warns major UN report

The IPCC synthesis report sets out how the world can stick to its 1.5°C, but says that countries are substantially off track to meet it.

By Daniel Capurro, Environment Correspondent

The world has the money and the means to tackle climate change but still isn't doing so, UN scientists have said after the publication of a landmark report.

Average global temperatures have already increased by 1.1°C and much of the warming for the next decade or so is already locked in because of ongoing emissions, while the impacts on sea levels are set to persist for thousands of years, according to the Intergovernmental Panel on Climate Change (IPCC).

As it stands, countries are well off course to meet the pledges they made at the Paris climate conference in 2015, when the 1.5°C target was agreed. Under currently implemented policies, the world will warm between 2.2 and 3.5 °C

Dr Hoesung Lee, the chair of the IPCC, said the world was 'walking when we should be sprinting'.

Dr Peter Thorne, an IPPC author, said the world would probably reach 1.5°C in 'the first half of the next decade' but that it was in our own hands whether we would 'go blasting through' and carry on beyond 2°C or whether there would be a small overshoot that could be recovered from.

'The future really is in our hands,' he said.

The report is clear that the technology and money are available to halt global warming.

'There is sufficient global capital to rapidly reduce greenhouse gas emissions if existing barriers are reduced,' the IPCC said on publication of the report.

However, the report points out that the public and private financial flows in fossil fuels still exceed those for clean energy.

'The most important message is that we do have the solutions. We do know how to do transformative adaptation. We do know how to incentivize the emission reduction. And that, I think, is the most important message,' said Dr Friederike Otto, a climate scientist at Imperial College and an author of the report.

'It's not that we lack some technology or that we lack knowledge. The thing that we are lacking so far globally is the sense of urgency'.

The report emphasises that developing countries are bearing the biggest costs of climate change having made the smallest contribution to it.

Antonio Guterres, the UN secretary general, responded by calling for rich countries to reach net zero ten years earlier than the rest of the world. Britain is currently aiming to hit net zero by 2050.

If that urgency does not arise, the report warns, the costs will only increase while more human and natural systems will cease to be viable.

For example, above 1.7°C of warming, much of the tropics would face more than 50 days a year where temperatures are too hot for humans to survive.

The IPCC published its sixth synthesis report on Monday after a week-long session by scientists working in Switzerland.

It will be the last major report from the IPCC for several years, and when the world next hears from the UN body it will likely be clear whether the world has failed to limit global warming to 1.5°C.

Reaching the 1.5°C target remains a monumental task. The IPCC says that emissions need to already be falling and to be cut in half by 2030. Last year, global emissions rose again to another record high, although the International Energy Agency said there were signs that they were finally plateauing.

On the other hand, the IPCC notes that the costs of renewable technologies such as wind turbines, solar panels and batteries have plunged over the last decade making them both economically and environmentally desirable.

The majority of the IPCC publication, as its name suggests, is a synthesis of the work done by the IPCC over its most recent cycle, which began in 2015.

It contains relatively little new science and is instead intended as a framework on which policymakers across the world can base their decisions.

Nevertheless, in that eight-year period, scientists have developed a firmer understanding of the consequences of climate change.

There is a greater concern about tipping points, beyond which climate change would rapidly accelerate. Once global warming passes 2°C, the IPCC warns that the Earth will reach a point where carbon sinks such as oceans and forests will lose their ability to sequester emissions.

The report also accepts that to the world may overshoot the 1.5°C target and then rely on removing carbon from the atmosphere, either through trees or technology, to return to that level before the end of the century.

Oliver Geden of the German Institute for International and Security Affairs and another author of the report said the issue would become significantly more important in the coming years.

'It will probably be one of the major issues in the next assessment cycle. So it's also building a bridge into the coming IPCC report,' he said.

20 March 2023

We can't stop climate change, but we can (still) avoid the worst-case scenario

By Christian Morris

Life on Earth is, in itself, a miracle. The conditions produced by the delicate balance of Earth's planetary systems has allowed modern humans to flourish and thrive over thousands of years. But, while we relish in the gifts this planet has given us, the fragile systems that sustain us – and all other life on this planet – are in peril.

Right now, the Earth is the hottest it's been in 12,000 years as a direct result of anthropogenic climate change. Humans have gravely altered our climate through the burning of fossil fuels and the degradation of ecosystems – such as rainforests, wetlands, and oceans – that have an innate ability to mitigate excess planet-warming gases.

While a shift in our climate is all but guaranteed due to our history of polluting the atmosphere, the latest research indicates that humans can still control just how bad the climate crisis gets. As of today, the concentration of carbon dioxide in the atmosphere sits at 413 parts per million (ppm), which is the highest level it's been in 800,000 years. With the concentration so high and the climate already having warmed a full degree Celsius, some impacts of climate change are inevitable. We're already seeing these impacts materialize; 2020 was tied as the warmest year on record and the U.S. experienced over 20 climate-fueled disasters, each totaling over $1 billion in damages.

Earth could turn into a 'hothouse.' What does that mean and what can we do about it?

We still can stabilize the climate

In some corners of the scientific community, climate change is viewed as an issue that is beyond the point of no return. Since carbon dioxide can remain suspended in the atmosphere for decades, some scientists have concluded that concentrations of the planet-warming gas would continue to warm the climate long after emissions were halted.

New research, however, refutes those claims and points out that there is hope. We can stabilize the climate if emissions drop to net-zero – meaning the remaining human-caused emissions balance out by removing the emissions from the atmosphere – which would limit warming and return the climate to a more steady state of being before we cross a tipping point. If humans are able to stop burning fossil fuels and emitting greenhouse gases, then the climate could level out in as little as ten years, as opposed to the previously held notion that it could take up to 20–25 years for stabilization.

What scientists had failed to account for in previous modelling was the influence that ecological dynamics have on atmospheric carbon dioxide concentrations, which naturally work to remove the gas. This includes the role nature plays in sequestering carbon, such as how wetlands, for example, are able to sequester and store massive amounts of carbon within their soils.

In an article from *InsideClimate News*, climate scientist Joeri Rogelj, a lead author on the upcoming IPCC report, remarks, 'It is our best understanding that, if we bring down CO2 to net zero, the warming will level off. The climate will stabilize within a decade or two […] There will be very little to no additional warming. Our best estimate is zero.'

Over 100 countries around the world have already committed to achieving net-zero emissions, and under President Biden, the U.S. recently committed to net-zero emissions by no later than 2050. For the second largest contributor to climate change, that's a good sign.

Dr. Michael Mann, a renowned climate scientist, comments on this revised scientific positioning on climate change in The Guardian, stating 'What this really means is that our actions have a direct and immediate impact on surface warming. It grants us agency, which is part of why it is so important to communicate this current best scientific understanding.'

'This is not a science problem anymore, it's an everything problem.'

What happens next is up to us

The big takeaway from these developments in climate science is that climate change may not be the end of life as we know it. We're not necessarily doomed. There is, in fact, still time to create a future where our planet remains livable for future generations, a planet that continues to sustain prosperous and healthy societies. Ultimately though, it's up to us to put in the legwork to create that world in the present because the lives and security of future generations are, quite literally, in our hands now.

With a change in leadership here in the U.S., we've seen President Biden take monumental steps towards achieving net-zero emissions and establishing the national framework we so desperately need to fight climate change from every facet of society. From stopping fossil fuel subsidies and mandating clean energy procurement, to centering climate change in both national and foreign policy, we can expect to see serious change as we help set the tone for global climate action.

Although the research provides a newfound hope that we can reverse course on climate change, it doesn't give us any more time to wait. To even begin reversing the climate crisis, we'll need to cut emissions in half by 2030 in order to achieve net-zero and begin removing heat-trapping gases from the atmosphere by 2050, and that effort will need to come from a breadth of policy solutions to aid in this transformative shift. We'll need to ramp up clean energy; implement carbon pricing programs; transition to all electric vehicle fleets; invest in reforestation, ecosystem restoration, and preservation; and so many more solutions rooted in equity and justice that will advance a sustainable future for our world. While hope is not lost, we've been given the opportunity to seriously act today to create a livable future for tomorrow.

4 February 2021

Climate change isn't a death sentence for the human race – demoralizing Gen Z isn't the answer

Activism is important, but it has to be done in a way that doesn't instill hopelessness in everyone around us.

By Danielle Butcher

In 1970, a group of college students created a day to celebrate and fight for a most worthy cause – our planet. Fifty-two years later, a new generation of young people are struggling beneath the weight of the environmental challenges we face, which impact nearly every facet of their lives. As the most activism-driven generation our country has ever seen, Gen Z is laser-focused on climate change. For many young folks, climate change dominates their thoughts and actions.

While their antics may seem silly, this constant activism has taken its toll on their mental health and outlook on life. It may be tempting to scoff at the notion, but statistics increasingly show a correlation between perceived environmental health and the real mental health of younger generations.

To put it simply, the kids are not all right.

Data from the American Psychological Association shows that 58 percent of American teens feel stressed out by climate change, while almost half of young people across the world have climate anxiety that affects their day-to-day life. Recently, a striking majority of OkCupid users noted that climate denial was their top deal-breaker for dating, and one-fourth of childless Americans say that climate change factored into their decision not to procreate. Finally, The Atlantic reports that 44 percent of American high school students experience 'persistent feelings of sadness or hopelessness' – the highest level of teenage sadness ever recorded, no doubt compounded by loneliness and stress.

But, why? Gen Z grew up in a world in which the media bombarded them with the message that climate change was on track to steal away their futures. Since they were children, millennials and Gen Z have been inundated with images of sad polar bears on melting ice caps, vast stretches of forests on fire, and animals soaked in oil following the infamous BP spill. Mainstream climate leaders like Al Gore and John Kerry cherry-picked worst-case climate scenarios while the climate movement railed against corporations and government, insisting that only systemic change could save us and that individual actions amount to nothing. When confronted with this news, young people have ended up depressed rather than inspired.

The most significant issues of our time have been communicated to young people in a way that hasn't instilled a sense of responsibility to act, but instead, a feeling of imminent doom and hopelessness. Climate change isn't a death sentence for the human race, and society has made a grave error by convincing younger generations that it is. Of course, climate change is a severe challenge that threatens our future, but we shouldn't be without hope this Earth Day or any other.

The environment is a crucial issue for my generation, and it's not red or blue. It's part of who we are – an inherent value. That's why it's become a deal-breaker in dating and why the political conversation around climate change has grown so heated. Climate denial isn't just a policy disagreement for Gen Z; it's a dismissal of a deeply-rooted fear.

Rather than shrugging or laughing these fears off, the generation that created Earth Day should remind today's young people that there is still cause for celebration. For instance, the regenerative farming efforts highlighted by Kiss the Ground are already in use by farmers all over the world and have massive carbon sequestration potential. Plans for small modular nuclear reactors are being submitted for federal approval and could revolutionize the way we view clean energy. The American bison, once almost extinct, now once again roams the western plains.

In all the talk of societal change and revolution, we've forgotten about the solutions we can implement today and the larger ones we can begin to work toward. By convincing young people that there's no future to look forward to, we've taken away the desire for actionable solutions that make a difference. Earth Day is the ideal time to remind ourselves that we should be solution-oriented when it comes to helping our planet, not just reactionary and scared.

22 April 2022

Generational divide over climate action a myth, study finds

Seven in 10 of all generations surveyed recognise the need for action.

Older people are just as likely as younger people to recognise the need for action on climate change and to say they're willing to make big sacrifices to protect the environment, suggesting claims of a generational divide over the future of the planet may be exaggerated, according to a new UK study marking the publication of the book Generations by Professor Bobby Duffy.

The research, by the Policy Institute at King's College London and New Scientist magazine, finds that around seven in 10 people from all generations surveyed say climate change, biodiversity loss and other environmental issues are big enough problems that they justify significant changes to people's lifestyles, with no real difference in agreement between Baby Boomers (74%) – the oldest generation polled – and Gen Z (71%), the youngest.

Similarly, there are almost identical levels of agreement across the generations that people themselves are willing to make big changes to their own lifestyle to reduce the impact of climate change: there is virtually no difference between the proportion of Baby Boomers (68%), Gen X (66%), Millennials (65%) and Gen Z (70%) who say they're prepared to make such a sacrifice.

Where there is some generational difference in views is on whether environmental concerns should take precedence over economic growth: 66% of Gen Z and 57% of Millennials agree environmental concerns should take priority over the economy, compared with 44% of Baby Boomers and 45% of Gen X.

But despite this, older generations are still more likely to agree than disagree that the environment should come first – for example, 24% of Baby Boomers think we shouldn't prioritise climate change over the economic growth, far lower than the 44% who think we should.

Younger people are more likely to be fatalistic about climate change

While younger people are often thought to be most active on climate issues, they are actually more likely than older generations to say there's no point in changing their behaviour to tackle climate change because it won't make any difference anyway: 33% of Gen Z and 32% of Millennials feel this way, compared with 22% of Gen X and 19% of Baby Boomers.

There is an even bigger gap between different generations when it comes to rejection of this idea: 61% of UK Baby Boomers disagree that there's no point altering their behaviour – compared with 41% of Millennials.

But the public perception is that older people are most likely to think changing their behaviour is pointless

Half the UK public (wrongly) believe that older people are most resigned about what they can do to save the environment.

49% think Baby Boomers and those in older generations are most likely to say there's no point changing their behaviour to tackle climate change, compared with 30% who think Gen X, Millennials and Gen Z are most inclined to feel this way. But the reality is that these younger generations are more likely to be fatalistic about this.

To what extent do you agree or disagree with the following statements? *Climate change, biodiversity loss and other environmental issues are big enough problems that they justify significant changes to people's lifestyles*

Baby Boomers **74%**

Gen X **69%**

Millenials **70%**

Gen Z **71%**

Source: Who cares about climate change? Attitudes across the generations

To what extent do you agree or disagree with the following statements? *I am willing to make significant changes to my own lifestyle to reduce the impact of climate change*

Baby Boomers **74%**

Gen X **69%**

Millenials **65%**

Gen Z **70%**

Source: Who cares about climate change? Attitudes across the generations

To what extent do you agree or disagree with the following statements? *There is no point in changing my behaviour to tackle climate change because it won't make any difference anyway*

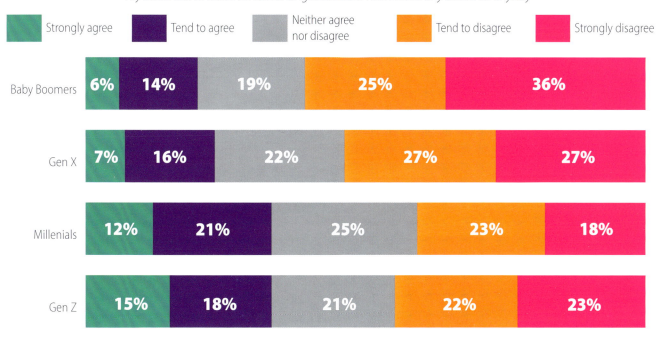

Strongly agree ・ Tend to agree ・ Neither agree nor disagree ・ Tend to disagree ・ Strongly disagree

	Strongly agree	Tend to agree	Neither agree nor disagree	Tend to disagree	Strongly disagree
Baby Boomers	6%	14%	19%	25%	36%
Gen X	7%	16%	22%	27%	27%
Millenials	12%	21%	25%	23%	18%
Gen Z	15%	18%	21%	22%	23%

Source: Who cares about climate change? Attitudes across the generations

Thinking about the UK population overall, on average, which of the following age groups do you think is most likely to say that there is no point changing their behaviour to tackle climate change because it won't make any difference?

Who the UK public think are most likely to say there's no point changing their behaviour

Baby Boomers & older: **49%**

Gen X, Millenials & Gen Z: **30%**

Reality

21% of Baby Boomers and older say there's no point in changing their behaviour...

... compared with **29%** of Gen X, Millenials and Gen Z combined

Source: Who cares about climate change? Attitudes across the generations

Thinking about the UK population overall, on average, which of the following age groups do you think is most likely to say that there is no point changing their behaviour to tackle climate change because it won't make any difference?

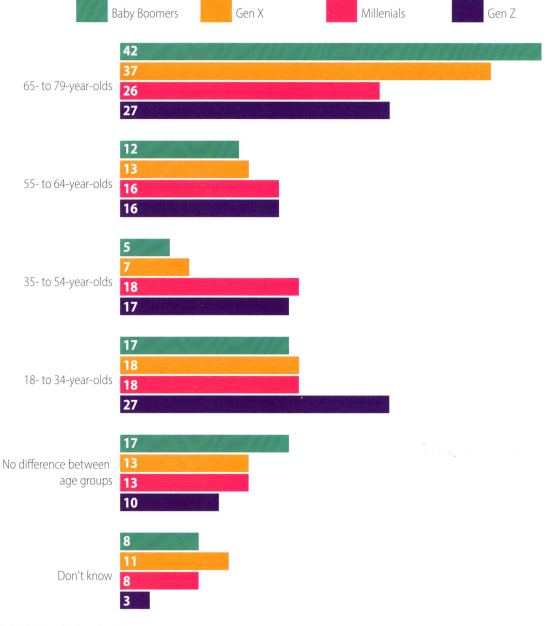

Baby Boomers — Gen X — Millenials — Gen Z

65- to 79-year-olds
- 42
- 37
- 26
- 27

55- to 64-year-olds
- 12
- 13
- 16
- 16

35- to 54-year-olds
- 5
- 7
- 18
- 17

18- to 34-year-olds
- 17
- 18
- 18
- 27

No difference between age groups
- 17
- 13
- 13
- 10

Don't know
- 8
- 11
- 8
- 3

Source: Who cares about climate change? Attitudes across the generations

And the public don't realise that 'cancelling' socially irresponsible brands is more of a middle-age thing

The public think that younger generations are most likely to have boycotted certain products for socially conscious reasons in the last year, with 27% guessing that Gen Z have done so and 23% saying the same about Millennials – much higher than the proportions who guess that Gen X (9%) and Baby Boomers (8%) have done so.

But according to previous research conducted as part of the European Social Survey, it is actually older generations who are most likely to have carried out such boycotts: for example, in 2018, 31% of UK Baby Boomers said they had boycotted a product as a way to improve things or prevent things going wrong – more than double the 12% of Gen Z who reported doing so.

Professor Bobby Duffy, director of the Policy Institute at King's College London, said:

'There are many myths about the differences between generations – but none are more destructive than the claim that it's only the young who care about climate change. When Time magazine named Greta Thunberg their person of the year in 2019, they called her a 'standard bearer in generational battle', which is reflective of the unthinking ageism that has crept into some portrayals of the environmental movement. But, as I examine in my new book, Generations, these stereotypes collapse when we look at the evidence.

'There is virtually no difference in views between generations on the importance of climate action, and all say they are willing to make big sacrifices to achieve this. What's more, older people are actually less likely than the young to feel that it's pointless to act in environmentally conscious ways because it won't make a difference. Parents and grandparents care deeply about the legacy they're leaving for their children and grandchildren – not just their house or jewellery, but the state of the planet. If we want a greener future, we need to act together, uniting the generations, rather than trying to drive an imagined wedge between them.'

Richard Webb, executive editor of New Scientist, said:

'There's been a lot of talk about the attitude of different generations towards the pressing issues of the day, not least the existential challenge of climate change and other aspects of our impact on the planet, but there's precious little in the way of hard data. At New Scientist we're all about informed debate, which was why we were pleased to join forces with Bobby and his team to get some facts on the table.

'The findings of the survey provide food for thought for policymakers ahead of the crucial COP26 climate summit in Glasgow in November. Far from being an obsession of a young, activist few, support for measures that put our lives on a more sustainable footing as we look to building back from the covid-19 pandemic command broad support across generations. They could be a route to increased engagement among groups increasingly disillusioned with politics.'

More information
Savanta ComRes surveyed 2,050 UK adults aged 18+ online between 2 and 9 August 2021. Data were weighted to be representative of UK adults by age, gender, region and social grade.

15 September 2021

www.kcl.ac.uk

Climate report: 'It's now or never, if we want to limit global warming to 1.5°C'

The window to stop dangerous climate change is closing, but solutions to slash emissions in the next decade are here now, say IPCC experts.

By Lottie Morgan, Siobhan Stack-Maddox, Nicole Kuchapski, Katrine Petersen, Hana Amer

The UN Intergovernmental Panel on Climate Change (IPCC)'s latest Working Group III Sixth Assessment Report, *Climate Change 2022: Mitigation of Climate Change*, focuses on solutions to climate change – what can be done to stop emissions rising.

It shows that levels of greenhouse gas emissions are now the highest they have ever been and that, without deep and immediate cuts in emissions, limiting global warming to 1.5°C – as agreed in the Paris Agreement – will be beyond reach.

Imperial's Professor Jim Skea is Co-Chair of Working Group III, the group of 278 authors responsible for the report. 'It's now or never, if we want to limit global warming to 1.5°C,' he said. 'Without immediate and deep emissions reductions across all sectors, it will be impossible.'

Current policies implemented to tackle climate change put the world on course for 3.2°C of warming, which would be devastating for people and ecosystems across the world.

'The latest report from the IPCC shows that, while we've made progress towards climate goals, there is still a huge gap between current action and the actions needed to limit warming to below more damaging levels,' said Dr Robin Lamboll, Research Associate at Imperial's Centre for Environmental Policy.

However, the report makes clear that technological, financial and governance tools to cut emissions in all sectors already exist, and many of these options are now cheaper and more feasible than ever – and can improve people's health and quality of life all over the world.

Climate action is a good investment

The report highlighted that the costs of solar energy and lithium-ion batteries have fallen by up to 85% in the past decade, and the cost of wind energy up to 55%, due to policies supporting innovation and enabling wider roll-out of the solutions.

For Michael Wilkins, Executive Director and Professor of Practice at Imperial's Centre for Climate Finance and Investment, these falling costs demonstrate that, 'with the right policy incentives and economic frameworks, climate change mitigation can be financed at scale and relatively quickly'.

Dr Ajay Gambhir, Senior Research Fellow at Imperial's Grantham Institute, echoed this, emphasising that, 'the falling cost of low-carbon technologies like renewables and electric vehicles will help us to decarbonise at low costs – just a few percentage points of our GDP by 2050'.

Dr Alex Köberle, Advanced Research Fellow at the Grantham Institute and contributing author to the report, further emphasised that, 'these GDP losses of "a few percent" are only telling half the story, since they do not account for the main benefit of mitigation action, that is, the avoided damages from climate change and lower adaptation costs', which could be very high.

Ambitious action on climate change also reduces the risk of triggering social and environmental tipping points at higher temperatures of 2°C and above, which would result in irreversible impacts. Furthermore, these costs do not account for the significant co-benefits that tackling climate change brings for health and wellbeing.

'The IPCC report not only shows that a lower-carbon future is technically and economically feasible, but that it is also a future to aspire to – with cleaner air, homes that everyone can afford to heat and greater energy security,' said Dr Neil Jennings, Partnership Development Manager at the Grantham Institute.

Richard Green, Professor of Sustainable Energy Business at Imperial, also highlighted the economic case for urgent climate action. He said: 'The report shows that early action is better for the economy than waiting before cutting emissions, and that while some of the things we need to do will be expensive, other options for cutting emissions will actually save us money – in the way that the most recent UK renewables are cheaper than the price of gas-fired electricity.'

Leaving fossil fuels in the ground

In contrast to the ambiguous wording on 'phasing down' fossil fuels in the Glasgow Pact after the UN Climate Summit (COP26) last November, the latest IPCC report calls for 'a substantial reduction in fossil fuel use', said Krista Halttunen, Research Postgraduate on the Science and Solutions for a Changing Planet Doctoral Training Partnership.

'Today, there is still more financing for fossil fuels than climate change adaptation and mitigation globally,' explained Krista, whose research focuses on the future of fossil fuels in the low-carbon energy transition.

'Emissions from existing and planned fossil fuel infrastructure today are already enough to take the world beyond 1.5°C of warming. This means that, to meet the Paris Agreement climate goals, many existing fossil fuel power stations need to be retired early or supplied with emission-reducing technologies,' she added.

Creating change in transport

Emissions in the global transport sector are still growing by about 2% each year. To reduce these, the IPCC proposes reducing demand for transport by encouraging a shift to more walking, cycling and public transport, as well as re-designing cities to support a move away from cars.

Imperial's Dr Drew Pearce, co-author of Research Pathways for Net Zero Transport, said: 'Active transport, public transport and demand reduction are policy "slam dunks": saving us money and emissions as well as leading to a slew of other health and wellbeing co-benefits'.

He continued: 'Cities and towns will play a crucial role in creating change through the whole supply chain. This too aligns with our recent work, still in review, which shows that significant modal shifts away from private cars are needed at a city-level and that there are significant emissions impacts outside of the city in the wider supply chain.'

Zero-carbon buildings

To reduce emissions in the building sector, the IPCC emphasises the need to increase energy efficiency and switch to electric solutions for heating powered by renewables. Ambitious schemes for retrofitting buildings can also improve wellbeing as well as help people to cope with the impacts of climate change, such as flooding or overheating.

Dr Richard Hanna, Research Associate at Imperial's Centre for Environmental Policy, said: 'Reducing emissions from energy use in buildings is an urgent priority... The current decade is critical for developing the required skills and supply chains. Policy packages to improve buildings' energy efficiency and replace high carbon heating with renewable heating need to be sufficiently ambitious, well-designed and effectively implemented.'

'I'm glad to see the acknowledgement that energy efficiency policies can contribute to 42% of the decarbonisation of the global building stock, with particular mention in the report of the skills and supply chains required to achieve this,' added Dr Kate Simpson, Research Associate in Housing Adaptation and Retrofit at Imperial's School of Design Engineering.

The role of carbon dioxide removal

This is the first IPCC Assessment Report that explores the full spectrum of carbon dioxide removal (CDR) approaches, and the opportunities, challenges and risks it poses. CDR involves taking carbon out of the atmosphere by, for example, planting trees or sucking carbon out of the air with machines and storing it underground.

'It might sound crazy to talk about carbon removal when we are still putting it up there at record rates. But the reality is that, if we're serious about pursuing a robust and affordable set of actions to get to net zero by 2050, then carbon removal is in the mix. And that means starting now,' said Dr Steve Smith, Executive Director at CO₂RE and Visiting Researcher at the Grantham Institute.

However, Dr Smith emphasised the importance of reducing emissions at source across all sectors as the first and most urgent step, rather than relying on CDR: 'With carbon removal getting lots of the headlines, it's crucial to remember we won't stop climate change without all the other stuff in the IPCC report: cutting energy waste, cutting fossil fuels if the carbon isn't captured and stored, better diets, and not cutting down forests,' he said.

The power of behaviour change

For the first time in an IPCC assessment report, there is a whole chapter dedicated to the role that behaviour change can play in bringing down emissions. The report shows that policies to reduce demand could cut emissions by 40-70% by 2050.

'Behavioural and lifestyle changes can reduce our emissions considerably, as will action in cities and other urban areas; and there are huge co-benefits to climate action, including to our health and wellbeing,' said Dr Ajay Gambhir.

'This is all encouraging, and means we have no excuse not to make immediate, sustained and deep reductions to our emissions, towards net-zero carbon dioxide in the early 2050s.'

Demand-side changes include encouraging people to switch to more sustainable diets, shifts to walking, cycling and more public transport, reducing waste and using products for longer.

6 April 2022

Why supporting women and girls is an overlooked climate solution

The Egyptian COP27 Presidency chose gender, alongside water, as the theme for proceedings on Monday (14 November). Here, we explain why gender equality is so important to delivering international goals on climate mitigation and adaptation.

By Sarah George

By now, you've likely seen the statistic from Care International that just seven of the 110 world leaders to have attended COP27 during the first week were women. If not, you'll have probably seen the memes of the group photos at the world leaders summit part of the conference on 7-8 November, bearing captions like 'maybe we should start asking the women'.

Women and girls account for 51% of the global population but only 21% of government ministers are women. The percentage drops even further for governmental positions heading up nations or states.

It will doubtless take decades to buck this trend. In the meantime, nations are increasingly recognizing that climate solutions and commitments need to properly include women in their design and implementation – especially given that women are more vulnerable than men to the physical impacts of climate change, mainly because they are more likely to live in poverty and to depend on at-risk natural resources.

The UN Framework Convention on Climate Change (UNFCCC) does host an agreement for nations to recognise that women and girls are disproportionately impacted by climate change and to link efforts relating to climate action and to gender equality. In a nutshell, nations state through the Gender Action Plan that they 'recognise that the full, meaningful and equal participation and leadership of women in all aspects of the UNFCCC process and in national and local-level climate policy and action is vital for achieving long-term climate goals'. They also acknowledge that 'gender-responsive implementation and means of implementation of climate policy and action can enable Parties to raise ambition.'

Care International is calling for the implementation of the Gender Action Plan to be given 'much greater priority' in this period in between updates to the text.

The NGO's gender equality expert Rosa Van Driel said: 'CARE calls on all countries to fully engage in emphasising gender-transformative climate action and decision-making throughout negotiations at COP27. This can be done by promoting women's and girls' leadership and participation in the negotiating rooms. Climate finance for mitigation, adaptation, loss and damage needs to be gender-transformative and reach grassroots and indigenous women and women-led and women's rights organisations involved in climate action.'

So, aside from the obvious statement that half of the global population should not be excluded from climate decisions which affect us all, why is it so important to involve women in climate leadership at all levels?

Family planning as a climate solution

Project Drawdown, which provides a list of climate solutions and advises policymakers and private sector entities regarding their implementation, has continually worked to highlight the importance of educating women and of improving family planning services to climate efforts.

It has estimated that, if ambitions and actions matched UN Sustainable Development Goals (SDGs) relating to these topics, 68.9 gigatons of CO_2e could be avoided between 2020 and 2050. For context, global CO_2e emissions in 2019 and 2020 totalled around 71.5 gigatons – so this is a significant saving, and from actions that also deliver social sustainability benefits.

The family planning piece's contribution to avoiding emissions is largely because it will slow population growth. The global population is projected to surpass eight billion people for the first time later this week, and the UN is also primed to confirm that India will surpass China as the most populous country in 2023.

It bears emphasising that the emissions footprint of an individual varies widely depending on where they live and their economic background. Someone from one of the world's 46 least developed countries will have a lower personal footprint, on average, than someone from the US. These 46 countries collectively account for 1.1% of total global annual emissions but 14% of the global population, according to the UN. In contrast, the US has about 4% of the global population and is responsible for around 28% of global annual carbon emissions. A report recently published by Oxfam states that billionaires generate one million times more CO2e over their lifetimes than those on average earnings.

Project Drawdown does not advocate that fertility should be limited or that nations should promote smaller families. Instead, it has stated, that those able to bear children should have 'full body autonomy' to decide whether to have children and how many, when, and with whom. The US Supreme Court's recent decision to overturn Roe v Wade has prompted more than a dozen states to tighten abortion ban or restriction rules, which goes against what Project Drawdown is advocating for.

Spotlight on education

Regarding education, Project Drawdown recommends that every person has access to at least 12 years of high-quality education. According to UNESCO, more than 58 million children of primary school age were not accessing basic education pre-pandemic, of which 32 million were female. Around half of these children are in conflict-afflicted regions and countries.

The problem is the most pronounced in Sub-Saharan Africa, where almost one in five children of this age group are missing school. Poverty is a key contributor to this trend – many districts cannot afford adequate school staff and supplies and families in this region may not be able to afford to send all children to school. More often than not, in this case, the boys are chosen to attend school while the girls stay home or go to work. In some cases, families feel they have no other choice than to get children to start work early, commonly in agriculture or manufacturing.

Project Drawdown emphasizes that, in this way, poorer children – and in particular, poorer girls – are excluded from education. Unicef has stated that only half of countries have achieved gender parity in primary education, dropping to 42% for lower secondary education and 24% in upper secondary education. As they grow older, these girls are left unable to participate in academic, commercial and advocacy spaces. In short, they have less of a chance to actively contribute to climate solutions.

There are several studies concluding that, because women in low-income nations traditionally act as stewards of household resources (such as food, fuel and water) and natural resources in their regions, there are benefits for climate when they are supported to apply these learnings on a broader scale. The UN has stated that women comprise 43% of the agriculture sector's workforce in the Global South and that women-led farms typically see crop yields that are 20-30% higher.

Similar research has been conducted to the need to ensure that women are properly involved in climate adaptation and disaster remediation work, given that they are more exposed to climate risks.

When we look at the participation of women in corporate leadership and policy leadership in more wealthy nations, there is also evidence that greater gender balance could bring better outcomes for the environment. One 2019 paper found that national parliaments with greater representation of women tended to adopt more ambitious climate policies.

In the private sector, a 2020 study from Bloomberg NEF concluded that having at least 30% women on a large business's board 'makes a key difference to climate governance and innovation'. The study analysed sustainability plans and emissions disclosure from 2,800 corporates. On average, corporates with boards consisting of 30% women or more averaged 0.6% emissions growth between 2016 and 2018. The increase was 3.5% within that timeframe for firms with no women on the board.

Spain's IE Business School has speculated that this trend could be down to the different ways women perceive morality and ethics, including their perceived responsibilities on helping others. Women, it has stated, are more likely to look at long-term stewardship rather than one-off and short-term interventions – and, of course, solving the climate and nature crises is a long-term effort.

14 November 2022

www.edie.net

Six lifestyle changes that could help avert the climate crisis

If everyone in the developed world rung in these changes, emissions would fall by a quarter, research suggests.

The sheer magnitude and complexity of the climate crisis can leave many people feeling helpless and apathetic. Why bother giving up beef if BP is drilling for more oil? What's the point in cutting back on air travel when airlines run empty flights just to keep their airport slots?

Anyone asking themselves such questions following the UN's latest warning about the climate crisis may be cheered by research published earlier this year. Commissioned by The Jump, a grassroots environmental group, it found that citizens have direct influence over 25-27 per cent of the emissions savings needed by 2030 to avoid climate chaos. In other words, people have more agency over the global heating than they might think.

That's not to absolve governments and corporations of their responsibilities, far from it. Without greater ambition from the public and private sectors, the climate crisis will intensify. This week's UN report – the most alarming to date – underscores the need for immediate and radical action to avoid the worst consequences of global heating. The window in which to act, it warned, is slamming shut.

The Jump's research was carried out by academics at Leeds University, in collaboration with the global engineering firm Arup and the C40 group of world cities. It found that making dietary changes is the single biggest thing people can do to reduce emissions, followed by giving up fast fashion.

'Citizen action really does add up,' said Rachel Huxley, director of knowledge and learning at C40 Cities. 'This analysis shows the collective impact that individuals, and individual choices and action, can contribute to combating climate change.'

The Jump was launched to help people in developed countries make lifestyle changes for the sake of the climate.

How can I help stop climate change

'The Jump is a fun grassroots movement of people leading the way to less stuff and more joy,' explained Tom Bailey, the movement's co-founder. 'Coming together to make practical changes, support and inspire each other, celebrate success and drive a shift in society's mindsets and cultures.'

The organisation has identified six lifestyle changes that people can make to directly reduce emissions. Those looking to take it further should check out the Positive News guide to taking climate action.

Six ways you can slash emissions, according to The Jump

1. Eat green

A shift to a mostly plant-based diet, combined with eliminating household food waste, would deliver 12 per cent of the total savings needed by North American and European countries.

2. Dress retro

Limiting yourself to buying three or fewer new items a year would deliver six per cent of the total savings needed. That means rummaging around in vintage shops more, and getting garments repaired, which you can do on most high streets or through apps like Sojo.

3. Shun the skies

Embracing low-carbon transport when going on holiday can reduce emissions by around two per cent. The Jump's research allows for a maximum of one short-haul flight every three years, and one long-haul every eight years.

4. Ditch the car

Reducing vehicle ownership (or, if possible, moving away from vehicle ownership altogether), would deliver two per cent of the total savings needed by 2030.

5. Keep hold of electronics

Extending the lifetime of electronics so they are used for at least seven years would deliver three per cent of the total savings needed. Helping people do that is the Restart Project, a charity that repairs broken electronics on a pay-what-you-can-afford basis. It has plans to open repair factories on every UK high street, and this week cut the ribbon on one such facility in London.

6. Change the system

To influence the remaining 73 per cent of emissions that are out of their direct control, citizens could take action that encourages and supports industry and government to make the high impact societal changes that are urgently needed.

For instance swapping to a green energy supplier, changing to a green pension, retrofitting our homes, or taking political action.

28 October 2022

The children of the revolution

One generation of world leaders has failed to halt climate change. The next generation intend to change all that.

By Suna Erdem

One of the unexpected delights of the Cop27 climate conference was the Indian activist Licypriya Kangujam's dogged questioning of Britain's climate minister, Zac Goldsmith, about the UK's jailed climate protesters. She chased him along corridors and through the doors that his entourage were trying to close in her face. Afterwards, she said, importantly: 'We need to hold lawmakers accountable for their political decisions.'

Kangujam is 11. She is part of the newest generation of climate activists who are putting adult politicians to shame with their straight talking and persistent demands, and they're getting louder and angrier as the world's temperatures reach catastrophic tipping points.

Amid the push for more action in the small window the world has left to avoid the worst, 2023 is set to be the year of youthful activists from the global south who, like Kangujam, are at the frontline of the climate emergency, in countries bearing the brunt of the changes they had almost no role in causing. Their dramatic, powerful stories of living through global warming-induced disasters, their savvy lobbying and increasing protest campaigns will make their mark after a year of yet more foot-dragging by national leaders determined to deprioritise armageddon.

Sweden's climate action wunderkind, Greta Thunberg, who has long dominated the limelight, has vowed to 'hand over the megaphone' to those in countries already suffering who have campaigned since their earliest days with a fraction of the publicity.

'I started almost as soon as I could speak. For us, it's a matter of survival,' said Ayisha Siddiqa, a prominent climate activist from Pakistan, now 23. Long under siege from rising global temperatures, in 2022 a third of Pakistan's land was engulfed by the worst floods in its history, following a severe heatwave. It cost more than 1,700 lives, affected another 33 million people, destroyed or damaged around 1.3 million homes and left 10 million children in immediate need of lifesaving support. 'Worlds came to an end,' Siddiqa said. 'My village doesn't even exist any more. This kind of disaster will happen again and again.'

But the damage began much earlier. As a small child, Siddiqa – now in the US after her father won a visa lottery – saw her own community 'condemned to multiple infrastructure projects, mining, dams,' all of which damaged the ecosystem and the livelihoods and health of those who lived there. The local river became dangerously polluted, affecting drinking water – she blames this for her grandparents' deaths. 'When the water they drank killed them, I knew I had to do something. I had to start speaking in public.'

There have been 'tears in the room' at Siddiqa's heartfelt speeches, but she emphasises that the world won't change through a 'public speaking tournament – panels, talks, yapping'. It needs money, action, and a refusal to bow to lobbyists, and the young campaigners need to work out how to play the negotiations, and get closer to forums where decisions are made.

Now, like Siddiqa – who campaigns for financial resources for the global south and fossil fuel non-proliferation – the new generation of campaigners such as Uganda's Vanessa Nakate, Iranian-American Sophia Kianni and Kenya's Eric Njuguna are knocking on the doors of political power. They were out in force at Cop27 in Egypt in November, some serving as part of the Youth Advisory Group on Climate Change, working with António Guterres. The United Nations secretary general is increasingly outspoken in his calls for leaders to act, saying that the world is 'on the highway to climate hell with our foot still on the accelerator', and that 'we are in the fight of our lives – and we are losing'.

'People often applaud António for his speeches. I can tell you that the reason the secretary general is so radical and so morally competent and headstrong is his young advisory group,' Siddiqa told me.

Younger activists can be different because they won't negotiate on vital issues, as they don't have the leeway and time to compromise that the older generations did. They ask for the stars, and hold firm.

One non-negotiable for them has been the 2015 Paris Agreement target of keeping global warming to 1.5C above pre-industrial levels – a figure chosen because beyond that is the extinction point of nations such as the Maldives, the Marshall Islands and Tuvalu. The world is currently around 1.2 degrees warmer, and is on course to wildly overshoot, so debates have begun on whether to set a new, more realistic target. But the young activists argue that if the target shifts, the urgency will reduce and the ultimate warming levels will probably be higher still. So far, it remains.

Another issue on which youth campaigners – together with indigenous activists – have persisted is the 'loss and damage fund', which was finally created at Cop27 to provide financial assistance to nations most vulnerable and affected by climate change. And it's from these nations that the most heart-rending stories emerge, bringing to life what has become an argument by numbers, of cynical trade-offs, of 'but, what about the economy?'

When Kangujam's escapades attract attention, she shifts focus to her own story, which reads like an odyssey through climate damage. Born in Manipur, a 'small carbon-negative state of India, full of rich biodiversity and alluring atmosphere', she went to school in Bhubaneswar, Odisha, where two large deadly cyclones hit in successive years. Delhi, supposedly safer, had dangerously high pollution levels and extreme heatwaves. India could soon become one of the first countries whose heatwaves exceed human survivability limits, according to a new World Bank study.

In 2019, aged six, Kangujam met world leaders, scientists, policymakers and other activists at the UN Disaster Conference in Mongolia, before founding the Child Movement – pushing for an Indian climate law, mandatory climate change education for all children, and for each of India's 350 million students to plant 10 trees.

Aged nine, she told an interviewer: 'As per historical data of the global carbon emissions, the global south is responsible for less than 10% of it but we're the biggest victim of global climate crisis today… We deserve our voices to be heard by the world, and the world leaders to act on it.'

She's very busy in 2023 – a schedule posted on her Twitter feed shows her attending 'town hall' meetings in Delhi, Goa, Dubai and Berlin in the first three months of the year, with 'school final exam' squeezed in between.

Kangujam and the other junior activists face adult lives of trying to survive and mitigate the effects of a crisis others did little to stop. They have had to grow up fast, getting to grips with complicated science, and learning how to brush off the usual backlash from those who object to being 'lectured' by schoolchildren, especially when the earnest, pint-sized activists – far from being naive and unrealistic – are more informed.

As Revan Ahmed, a 12-year-old from Libya who was the youngest Unicef delegate at Cop27, said at the conference: 'If you don't listen to us now, when will you hear us? We are living the disasters now. We, the children, will not be victims. We want to change the situation.'

It's an inspiring, but at the same time incredibly sad picture of young people working hard on both a local and international scale to push leaders of stature in the right direction, losing their innocence and childhood because of it, spurred on by the horrors they've seen.

Kianni, now 22, began her life in activism when at middle school. One night in Tehran the smog was so powerful she could not see the stars. Her Iranian relatives had no notion of the pollution caused by dirty energy, so she translated information for them. Later, she skipped school in the US to join Extinction Rebellion protests and held a hunger strike outside the offices of the then US House Speaker Nancy Pelosi. In 2020, she became the UN's youngest youth climate adviser. Shocked that the vast majority of scientific climate research – even the UN's definitive IPCC reports – was in languages that many people could not understand, she founded the youth-led NGO Climate Cardinals, which features translations into more than 100 languages by 8,000 student volunteers.

'Just six years ago, two out of five adults had never heard of climate change,' Kianni explained in a TED talk. '(They) deserve to have access to the resources they need to make sense of these disasters destroying their communities. The more people are informed about the climate crisis, the greater chance we have to coordinate collective efforts in protection of the future of our planet.'

Some of those most affected by climate change now – and increasingly vocal about it – are indigenous communities around the world, numbering around 370 million people globally, from the Ecuadorian Amazon's Sarayaku people to the Saami people living in northern Finland, Norway and Sweden, the Miskito community on Nicaragua's Caribbean Coast to the Baka tribe living in the rainforests of Cameroon.

Martina Fjällberg, a 22-year-old reindeer herder and environmental activist from Sweden's Saami community, is taking a degree in order to make scientists and politicians take her knowledge of nature more seriously. Swedish reindeer herders have long noticed the deterioration of nature – they now have little use for at least a quarter of their 200 words for snow as their lands warm – and work hard to sustain, defend and restore the biodiversity essential for their survival.

'Indigenous peoples' tangible experience of nature's decline gives them more insight than most into the threats facing our planet, so why are they still being forgotten in these debates?' she wrote in a joint article for Reuters with Princess Esmeralda of Belgium, an indigenous rights activist. '91% of (indigenous peoples') lands (and those of local communities) are considered to be in good or fair ecological condition….'

Despite the immediacy of climate change for the global south, the spotlight is still mostly on the global north which, even considering the drought and wildfires in Europe over the summer, is less affected.

When I speak to Eric Njuguna, the 20-year-old activist from Kenya, he points out that more people in the capitals of the world's richest countries have talked about campaigners throwing tomato soup on priceless artworks (protected by glass) than the thousands dying in Kenya and Somalia through ongoing droughts: 'The fact that stories from the global south are not being amplified has led to a huge disconnect between the reality on the ground – what the true reality of the climate crisis means,' he said, 'and those making decisions on issues such as mitigation finance playing down the urgency.'

I ask him what stories negotiators should be hearing now. 'Stories of mothers who have lost children in their arms because they haven't been able to feed them,' he said, without hesitation. 'Stories of the ongoing drought in the Horn of Africa… which is affecting communities with the least role in causing the crisis. They're bearing the brunt, losing families, losing the animals they depend on as a result of multiple failed rainy seasons.'

An activist since his early teens after severe droughts in Nairobi affected his school's water supply, Njuguna organised the group Zero Hour and then Fridays for the Future Kenya, inspired by Thunberg's school strikes. Last year he co-authored an essay with Thunberg and others for the New York Times, highlighting Unicef's 2021 report that found 2.2 billion children were at 'extremely high risk' of experiencing the consequences of climate change. He has met international politicians, including ministers. 'But each time talking to them I feel there's nothing new I can ever tell them – they have access to the best possible climate science, they have access to media sources, including the stories of journalists from here… But there's the fact that they still want to continue the fossil fuel era when the science is clear that we need to stop oil and gas exploration.'

A record 600 fossil fuel industry lobbyists were allowed at Cop27, in effect negotiating on behalf of corporations with governments to water down declarations. According to the financial think tank Carbon Tracker, Shell, TotalEnergies, Chevron and other oil majors recently approved $166bn (£138bn) of investment in new oil and gas fields, more than a third destined for sites that will only be needed if fossil fuel demand reaches levels that push warming above 2.5C.

Young activists I spoke to see 21st-century capitalism as a huge barrier to successful climate action, in the face of energy companies trying to extract just another couple of decades of profit before they must throw in the towel. Even new measures such as carbon credit-trading enable big polluters to buy the right to pollute from poorer nations, they complain, which are then left with little room to manoeuvre with their own emissions. As for the loss and damage fund, this is an 'empty bucket', they say, and even if eventually filled, the donors will preen about helping the poor – yet should understand they're really paying 'reparations, not aid'.

To push their point, however, they need to be at the table with power and a vote, as fully-fledged youth negotiators who will have to deal with the consequences of any failure.

'Of the world's population, 60% is below the age of 35. Young people are the greatest stakeholders,' says Siddiqa… 'Things will only change if we are helping write policies. I don't think this is impossible or romantic, or a pipe dream. It just needs political will.'

5 January 2023

9 things you can do about climate change

As experts on climate change, many people ask us, what can I do personally about it?

And how does this fit into the bigger picture?

We spoke to our scientists and drew up a list of the most achievable ways you personally can make a difference. While individuals alone may not be able to make drastic emissions cuts that limit climate change to acceptable levels, personal action is essential to raise the importance of issues to policymakers and businesses.

Using your voice as a consumer, a customer, a member of the electorate and an active citizen, will lead to changes on a much grander scale.

'Use your voice, use your vote, use your choice' - Al Gore

1. Make your voice heard by those in power

Tell your Member of Parliament, local councillors and city mayors that you think action on climate change is important.

A prosperous future for the United Kingdom depends on their decisions about the environment, green spaces, roads, cycling infrastructure, waste and recycling, air quality and energy efficient homes.

Ultimately, steps to reduce carbon emissions will have a positive impact on other local issues, like improving air quality and public health, creating jobs and reducing inequality.

What can I do?

Find out who your MP is, and the best way to contact them.

Join a social movement or campaign that focuses on environmental activities or gets everyone talking about climate change action, such as the Youth Strike 4 Climate or Extinction Rebellion.

There are many benefits to taking action on climate change, such as improved health, growth in the low-carbon jobs market, and reduced inequality.

2. Eat less meat and dairy

Avoiding meat and dairy products is one of the biggest ways to reduce your environmental impact on the planet. Studies suggest that a high-fibre, plant-based diet is also better for your health - so it can be a win-win.

Eat fewer or smaller portions of meat, especially red meat, which has the largest environmental impact, and reduce dairy products or switch them for non-dairy alternatives .

Try to choose fresh, seasonal produce that is grown locally to help reduce the carbon emissions from transportation, preservation and prolonged refrigeration.

The carbon footprint of one cheeseburger is equivalent to nine falafel in pitta; and six fish and chips

Find out more

Have you thought about eating insects? They are a healthy and environmentally friendly food source, so why don't we eat them? Researchers at Imperial have been investigating how people in the Western world can be convinced to eat them.

The carbon footprint of one cheeseburger is equivalent to...

9x falafel & pitta   **6x** fish & chips

3. Cut back on flying

If you need to fly for work, consider using video-conferencing instead. For trips in the same country or continent, take the train or explore options using an electric car.

When flying is unavoidable, pay a little extra for carbon offsetting.

For leisure trips, choose nearby destinations, and fly economy – on average, a passenger in business class has a carbon footprint three times higher than someone in economy.

Find out more

How do the cost, time and carbon emissions of a single journey from London to Amsterdam by plane, compare to travelling by train? (Assumptions: the start and end location are the main train terminal in the centre of each city; travel costs are for the cheapest advance tickets bought in advance

Single journey from London to Amsterdam

Plane
cost: £54
time: 4hrs 47min
CO_2: 58kg

Train
cost: £35
time: 4hrs 30min
CO_2: 3kg

and include all connecting journeys; carbon emissions are calculated using UK government greenhouse gas emissions factors for short-haul flights, ferry and international rail travel.)

There are a variety of reputable carbon offsetting schemes that fund sustainable development projects or natural solutions like planting trees.

Myclimate.org also compares the carbon emissions of your particular flight, with the maximum amount of carbon dioxide a person should produce per year in order to halt climate change, and the average amount an EU citizen produces each year. It makes for sobering reading.

Did you know?

Transport has become the largest emitting sector of the UK economy, accounting for 28% of UK greenhouse gas emissions in 2017.

4. Leave the car at home

Instead of getting in the car, walk or cycle – and enjoy the physical and mental health benefits, and the money saved. For longer journeys, use public transport, or try car sharing schemes.

Not only do cars contribute to greenhouse gas emissions, but air pollution caused by exhaust fumes from traffic poses a serious threat to public health. It has been shown to affect the health of unborn babies and increase the risk of dementia.

Furthermore, Imperial research shows that poor air quality in the capital leads to around 1,000 London hospital admissions for asthma and serious lung conditions every year, and that air pollution in the United States is associated with 30,000 deaths and reduced life expectancy.

If driving is unavoidable…

Investigate trading in your diesel or petrol car for an electric or hybrid model. Alternatively, if you only need one for a short time, there are some all-electric car hire companies.

When behind the wheel, think about the way you drive:

Switch off the engine when you park up.

Make sure the tyres are fully pumped, and that the oxygen sensors are in good order – this can improve the car's fuel mileage and efficiency by up to 3% and 40% respectively.

Drive smoothly.

5. Reduce your energy use, and bills

Small changes to your behaviour at home will help you use less energy, cutting your carbon footprint and your energy bills:

- Put on an extra layer and turn down the heating a degree or two.
- Turn off lights and appliances when you don't need them.
- Replace light bulbs with LEDs or other low-energy lights.
- Make simple changes to how you use hot water, like buying a water-efficient shower head.

Go further

Make sure your home is energy efficient. Check the building has proper insulation, and consider draught-proofing windows and doors. If you are in rented accommodation, lobby your landlord to make sure the property is energy efficient.

Switching energy supply to a green tariff is a great way to invest in renewable energy sources – and could save you money on bills too.

6. Respect and protect green spaces

Green spaces, such as parks and gardens, are important. They absorb carbon dioxide and are associated with lower levels of air pollution.

They help to regulate temperature by cooling overheated urban areas, can reduce flood risk by absorbing surface rainwater and can provide important habitats for a wide variety of insects, animals, birds and amphibians.

They also provide multiple benefits to public health, with studies linking green space to reduced levels of stress.

What can I do?

Plant trees. The Woodland Trust are aiming to plant 64 million trees over the next 10 years – and need your help. Whether you want to plant a single tree in your garden, or a whole wood, they have tools and resources to help.

Create your own green space. Add pot plants to your window sill or balcony, and if you have your own outdoor space, don't replace the grass with paving or artificial turf.

Help to protect and conserve green spaces like local parks, ponds or community gardens. Organisations like Fields In Trust and the National Federation of Parks and Green Spaces have advice and resources on how you can get involved in areas local to you.

Check out TCV. If you don't have direct access to open spaces, this community volunteering charity brings people together to connect to nature, and create healthier and happier communities.

7. Invest your money responsibly

Find out where your money goes. Voice your concerns about responsible investment by writing to your bank or pension provider, and ask if you can opt out of funds investing in fossil fuels.

There are also a number of 'ethical banks' you can investigate.

Find out more

Banks, pensions funds and big corporates often hold investments in fossil fuel companies. However, the discussion around responsible investment – weighing up environmental, social and governance (ESG) factors and taking them into consideration when investing money – is growing.

8. Cut consumption – and waste

Everything we use as consumers has a carbon footprint.

Avoid single-use items and fast fashion, and try not to buy more than you need.

Shop around for second-hand or quality items that last a long time.

Put your purchasing power to good use by choosing brands that align with your new green aspirations.

Try to minimise waste

Repair and reuse.

Give unwanted items a new life by donating them to charity or selling them on.

Avoid wasting food.

Let brands know if you think they are using too much packaging – some will take customer feedback seriously.

9. Talk about the changes you make

Conversations are a great way to spread big ideas.

As you make these positive changes to reduce your environmental impact, share your experience with your family, friends, customers and clients. Don't be a bore or confrontational. Instead, talk positively, and be honest about the ups and downs.

Find out more

For some tips on successful climate-based conversations, check out Climate Outreach's work with climate scientist and communicator Katherine Hayhoe.

For more on how we can achieve a cleaner, greener, fairer future, visit the Grantham Institute homepage.

The above information is reprinted with kind permission from Imperial College London.

© 2023 Imperial College London

www.imperial.ac.uk

Further Reading/ Useful Websites

Useful Websites

www.climate-xchange.org

www.edie.net

www.fairplanet.org

www.greenpeace.org.uk

www.greenpeace.org.uk

www.imperial.ac.uk

www.independent.co.uk

www.inews.co.uk

www.kcl.ac.uk

www.kids.frontiersin.org

www.parliament.uk

www.positive.news

www.quantamagazine.org

www.theneweuropean.co.uk

www.un.org

www.weforum.org

Page 1: From Climate Action, by United Nations, ©2023 United Nations. Reprinted with the permission of the United Nations. URL: https://www.un.org/en/climatechange/what-is-climate-change Date accessed: 16 February 2023

References for pages 16-19

[1] Marcott, S. A., Shakun, J. D., Clark, P. U., and Mix, A. C. 2013. A reconstruction of regional and global temperature for the past 11,300 years. Science. 339:1198–201. doi: 10.1126/science.1228026

[2] Available online at: https://climate.nasa.gov/vital-signs/global-temperature/

[3] Keeling, C. D., Bacastow, R. B., Bainbridge, A. E., Ekdahl, C. A., Guenther, P. R., and Waterman, L. S. 1976. Atmospheric carbon dioxide variations at Mauna Loa Observatory, Hawaii. Tellus. 28:538–51. doi: 10.3402/tellusa.v28i6.11322

[4] IPCC. 2021. "Summary for policymakers," in Climate Change 2021: The Physical Science Basis. Contribution of Working Group I to the Sixth Assessment Report of the Intergovernmental Panel on Climate Change, eds V. Masson-Delmotte, P. Zhai, A. Pirani, S. L. Connors, C. Péan, S. Berger, N. Caud, Y. Chen, L. Goldfarb, M. I. Gomis, M. Huang, K. Leitzell, E. Lonnoy, J. B. R. Matthews, T. K. Maycock, T. Waterfield, O. Yelekçi, R. Yu, and B. Zhou (Cambridge: Cambridge University Press).

[5] Trascasa-Castro, P., and Smith, C. 2021. What can we do to address climate change? Front. Young Minds. 9:672854. doi: 10.3389/frym.2021.672854

References for pages 20-23

[1] Marcott, S. A., Shakun, J. D., Clark, P. U., and Mix, A. C. 2013. A reconstruction of regional and global temperature for the past 11,300 years. Science. 339:1198–201. doi: 10.1126/science.1228026

[2] Available online at: https://climate.nasa.gov/vital-signs/global-temperature/

[3] Keeling, C. D., Bacastow, R. B., Bainbridge, A. E., Ekdahl, C. A., Guenther, P. R., and Waterman, L. S. 1976. Atmospheric carbon dioxide variations at Mauna Loa Observatory, Hawaii. Tellus. 28:538–51. doi: 10.3402/tellusa.v28i6.11322

[4] IPCC. 2021. 'Summary for policymakers,' in Climate Change 2021: The Physical Science Basis. Contribution of Working Group I to the Sixth Assessment Report of the Intergovernmental Panel on Climate Change, eds V. Masson-Delmotte, P. Zhai, A. Pirani, S. L. Connors, C. Péan, S. Berger, N. Caud, Y. Chen, L. Goldfarb, M. I. Gomis, M. Huang, K. Leitzell, E. Lonnoy, J. B. R. Matthews, T. K. Maycock, T. Waterfield, O. Yelekçi, R. Yu, and B. Zhou (Cambridge: Cambridge University Press).

[5] Trascasa-Castro, P., and Smith, C. 2021. What can we do to address climate change? Front. Young Minds. 9:672854. doi: 10.3389/frym.2021.672854

Air Pollution

Air pollution can cause both short-term and long-term effects on health and many people are concerned about pollution in the air that they breathe. These people may include people with heart or lung conditions, or other breathing problems, whose health may be affected by air pollution.

Carbon footprint

A carbon footprint is a measure of an individual's effect on the environment, taking into account all greenhouse gases that have been emitted for heating, lighting, transport, etc. throughout that individual's average day.

Carbon offsets

Carbon offsets are a reduction in greenhouse gas emissions made in order to compensate for greenhouse gas production somewhere else. Offsets can be purchased in order to comply with caps, such as the Kyoto Protocol. For example, rich industrialised countries may purchase carbon offsets from a developing country in order to satisfy environmental legislation.

Carbon Tax

Carbon tax is a fee imposed on the production, distribution and burning of fossil fuels responsible for CO2 and other greenhouse gas emissions. It incentivises companies to cut their carbon emissions and invest in cleaner, greener options.

Climate change

Climate change describes a global change in the balance of carbon absorbed and emitted into the atmosphere. This imbalance can be triggered by natural or human processes. It can cause either regional or global changes in weather averages and frequency of severe climatic events.

Climate models

Scientific models which are designed to replicate the Earth's climate. Scientists are able to hypothetically test the effects of global warming by simulating changes to the Earth's atmosphere.

CO2 emissions

Carbon dioxide gas released into the atmosphere. CO2 is released when fossil fuels are burnt. An increase in CO2 emissions due to human activity is arguably the main cause of global warming.

Conservation

Safeguarding biodiversity; attempting to protect endangered species and their habitats from destruction.

Eco-friendly

Policies, procedures, laws, goods or services that have a minimal or reduced impact on the environment.

Ecosystem

A system maintained by the interaction between different biological organisms within their physical environment, each one of which is important for the ecosystem to continue to function efficiently.

Emissions

Emissions are the exhaust fumes that are released from vehicles as they burn fuel. They are damaging to both the environment and people's health, containing, among other chemicals and greenhouse gases, carbon dioxide and carbon monoxide.

Energy

A force which powers or drives something. It is usually generated by burning a fuel such as coal or oil, or by harnessing natural heat or movement (for example by using a wind turbine).

Environment

The complex set of physical, geographic, biological, social, cultural and political conditions that surround an individual or organism and that ultimately determine its form and the nature of its survival.

Extinction Rebellion

Extinction Rebellion (abbreviated as XR) is a global movment of climate activists using civil disobedience and non-violent resistance to protest against climate breakdown and prevent mass extinction.

Fossil fuels

Fossil fuels are stores of energy formed from the remains of plants and animals that were alive millions of years ago. Coal, oil and gas are examples of fossil fuels. They are also known as non-renewable sources of energy, because they will eventually be used up: as they are finite, once they are gone we will be unable to produce more of them.

Global footprint

A person's global footprint refers to the impact that they have on the planet and the people around them, taking into account how much land and water each person needs to sustain their lifestyle.

Global warming

This refers to a rise in global average temperatures, caused by higher levels of greenhouse gases entering the atmosphere. Global warming is affecting the Earth in a number of ways, including melting the polar ice caps, which in turn is leading to rising sea levels.

Pollution

Toxic substances which are released into the environment: for example, harmful gases or chemicals deposited into the atmosphere or oceans. They can have a severe negative impact on the local environment, and in large quantities, on a global scale.

Sustainable Development Goals (SDGs)

17 goals set out by the United Nations to protect the planet and ensure that people around the world can live with equality and in a healthy environment by 2030. The goals cover social, economic and environmental sustainability. 'End poverty in all its forms everywhere' is the number one SDG.

Index

A
agriculture 3, 6, 18
air pollution 43
argon 4
asteroids 22

B
banks, ethical 41

C
carbon cycle 19
carbon dioxide 1, 3, 4, 6, 8, 17–19, 21, 23, 43
    removal 33
carbon footprint 39, 39–41, 43
carbon offsets 43
carbon tax 43
chlorofluorocarbons 3, 8
climate change
    causes 2–3, 6–7, 16–23
    definition 1, 5, 9
    effects/impacts of 3, 10–15
    generational views on 27–31, 37–39
    solutions 1, 3, 7, 24
    tipping points 7
    see also global warming; greenhouse gas emissions
climate refugees 1
conservation 43
COP27 34, 37

D
dairy products 39
decarbonisation 33
deforestation 6, 7
drought 11

E
Earth's orbit 2, 20–21
ecosystem 43
education 15, 35
emissions 43
    see also greenhouse gas emissions
energy 33, 43
    efficiency 40
    fossil fuels 1, 3, 6, 18–19, 43
    renewable 1, 33
environment 43
ethical banks 41
evolution 22–23
extinction 10, 22, 23, 37
Extinction Rebellion 38, 39, 43

F
F-gases 4
flooding 10–11
flying 36, 40
food production 6
fossil fuels 1, 3, 6, 18–19, 33, 43

G
global footprint 43
global warming 1–2, 5, 8–9, 10–11, 16–20, 24, 32, 43
    see also climate change
greenhouse gas emissions 1, 3–5, 6–7, 8, 19, 32
Greenpeace 7
green spaces 40–41

H
healthcare 15
heatwaves 11, 14–15
housing 15, 33

I
ice ages 2
ice sheets 7, 11
industrialisation 3, 6, 18
insects, as food 39
Intergovernmental Panel on Climate Change 5, 7, 32

L
landfills 1
lifestyle changes 36, 39–40

M
meat products 39
methane 1, 3, 6, 8, 19, 23
Milankovitch Cycle 2, 20

N
net zero emissions 1, 5, 26, 33
nitrogen 4
nitrous oxide 3, 4, 6, 19

O
oceans, and climate 3, 7
oxygen 4
ozone 3

P
Paris Agreement 1, 5, 7, 10, 32
permafrost 7
plastics 6
plate tectonics 21–22

polar ice sheets 7, 11
pollution 6, 43

R
rainforests 7
renewable energy 1, 33

S
saltwater intrusion 1
sea level rises 1, 13
solar cycles 20
solar radiation 2
Sun
    solar cycles 20
    solar radiation 2
    temperature 21

T
Thunberg, Greta 37
transport 6, 15, 33, 36, 40

U
United Nations
    Framework Convention on Climate Change 1, 34
    Sustainable Development Goals 1, 34, 43

V
volcanic eruptions 3
volcanic sulfur 20

W
waste 6, 41
weather 2, 9, 20
    extreme 10–11
    see also climate change
wildfires 11
women and girls, and climate action 34–35